973.13
C717t **Tornöe, Johannes Kristoffer,** 1891–
  Columbus in the Arctic? And the Vineland literature, by J. Kr. Tornöe. Oslo, 1965.
  92 p. maps (1 fold.) 23 cm.
  Bibliography: p. [85]–92.

  1. America—Disc. & explor.—Norse. 2. Columbus, Christopher. ɪ. Title.

E105.T69

500
65–8717

COLUMBUS IN THE ARCTIC?
AND
THE VINELAND LITERATURE

COLUMBUS IN THE ARCTIC?

AND

THE VINELAND LITERATURE

Part I
Notes on the Literature of the Vineland Voyages

Part II
Subsequent Voyages with Reference to Columbus

By

J. Kr. Tornöe

OSLO 1965

All rights reserved

By the author:
Grönlandssaken i parti-politikkens tjeneste, 1933
Hvitserk og Blåserk, Norsk Geografisk Tidsskrift, 1935
Lysstreif over Noregsveldets historie, 1944
Early American History: Norsemen before Columbus, 1964

Columbus in the Arctic?
Distribution:
Bokcentralen, Oslo 1, Norway

Printed in Norway by
A. W. Brøggers Boktrykkeri A/S
Oslo 1965

Contents

Introduction	7
Part I Notes on the Literature of the Vineland Voyages	9
Evidence from the Sagas	11
Evaluation of the Literature	14
The first histories	14
Another opinion – Gustav Storm	17
Grassland, not Vineland?	22
In northern mists	22
Objection	25
G. M. Gathorne-Hardy	26
Other opinions	30
A Leiv Eiriksson Memorial	32
Chronological Survey regarding geographical discoveries made from Iceland and Greenland	35
Part II Subsequent Voyages with Reference to Columbus	39
Columbus in the Arctic?	41
Greenland's Bond with the New World	43
Through the Catholic Church	43
From the Pope's letter	46
Newfoundland is an island	48
Colonization	49
The State of Greenland	52
Population	52
The need for resources	53
Working the new lands	54
Expansion to the new lands	56
Greenland in the 'Age of Exploration'	57
Scandinavian-Portuguese collaboration	58
A joint expedition	59
A Portuguese views the New World	60
Greenland is still a link to America	60
The Pining-Pothorst Expedition	63
Knowledge from the Norsemen	63
To India through the St. Lawrence	64
South, around Vineland	65
Another expedition	67
The Northwest Passage	69
The Johannes Skolp (Scolvus) expedition	70
The King's Mirror	71
The mystery of Columbus	73
Columbus at 73° north	74
Addendum	79
Bibliography	85

Introduction

It was when my work on Leiv Eiriksson and the Vineland voyages was almost completed that I first came across suggestions that Columbus well knew were he was going when he set out on what became known as his discovery of America. The suggestion was that he knew of Vineland and other parts of the American coast, and that he had acquired this information not only from Portuguese who had been on expeditions in the North, but from his own first-hand acquaintance with the Arctic.

It was the possibility of Columbus's own experience of the Arctic that led me to consider the state of Greenland at that time and to trace the possible thread of historical information which stretched from the time of the Vineland voyages to the time of Columbus. In this book, therefore, I present this possibility as a distinctly credible, though not proven, historical fact, and that is my main reason for beginning with an evaluation of the literature on Vineland, in order to clarify the question of the status of the sagas; for it is upon these that our historical interpretations are originally based.

The voyages of the early Vikings have long been subject of debate among historians, archaeologists, and those interested in establishing the course the journeys took and what lands they made accessible. The impact upon the records of history made by such controversies is momentous as more and more information is brought to light. We have only to consider the dating of the discovery of America to illustrate the historical revisions that are necessary as a result of the work of investigation.

Sifting through legends and documents, gathering all scraps of information, fitting together the pieces in a plausible manner to reconstruct the past events is an arduous task fraught with guesswork, controversy, doubt and scepticism; the more that is written about a subject the more complex that subject becomes, yet the increasing material affords us greater chances for arriving at truer conclusions.

A great deal has been written about the Viking voyages and much uncertainty has arisen over their supposed routes along the coast of North America. My attempts to clarify the picture have led me to the conclusions set out in my recent book, *Early American History: Norsemen before Columbus* (Universitetsforlaget, Oslo, 1964), and in this volume.

In *Early American History* I established, through correlation with the Viking sagas, the routes of the first Vikings to and along the North American coast and have indicated the evidence pointing to the existence of Leiv Eiriksson's settlement in Vineland about 1001 at what is now Falmouth, Massachusetts, not further north as is believed.

In this volume, then, I seek to support my views, and evaluate in the light of my evidence the various other viewpoints of the many scholars who have also sought to clarify the picture. The recent discovery of traces of Viking settlements at L'Anse aux Meadows in northern Newfoundland is of major interest in this connection.

Through my research briefly set forth here in my books, I hope to provide, not only a basis for interesting historical conjecture on Columbus, but also some small aid to those scholars who will consider all the literature more closely. I should like to mention also that for teachers of the recent developments and innovations in early American History I have prepared a handbook which will shortly be ready for publication. In all my work it has been my earnest hope to contribute to the clarification of the Vinland voyages and to trace their consequences for later times.

J. Kr. T

Oslo 1965

PART I

NOTES ON THE LITERATURE OF THE VINELAND VOYAGES

Evidence from the Sagas

In my previous book, *Early American History: Norsemen before Columbus*, I have traced the route of the first known Norseman to come to America, in 986, Bjarne Herjulfsson. Bjarne's ship, sailing from Iceland, missed its destination, the southern tip of Greenland and during bad weather conditions drifted far south into the Atlantic. When the fog lifted after many days, he sighted land. But this land did not fit the description of Greenland, and so without going ashore he sailed on northward, passing a second and a third land before reaching Greenland. I have reconstructed Bjarne's route after the fog lifted as having passed Nova Scotia, the first land he spotted, Newfoundland–Labrador, Bjarne's second land, and Baffin Land, his third land, before turning east to Greenland.

Leiv Eiriksson's expedition in 1001 I have accordingly indicated as tracing Bjarne's voyage in reverse, naming Bjarne's third land Helluland, his second land Markland, and his first land Vineland, this last being Nova Scotia and all the land southward that they knew of, and called Vineland from the grapes they found there. I have located his settlement, Leiv's *Budir* in Vineland, at the present Falmouth, Massachusetts.

Thorvald, Leiv's brother, later prepared to explore the size of Vineland. Sending one boat southward, he himself took a crew northward from Leiv's *Budir* and went from Falmouth to Cape Cod and then to Plymouth Harbor coming to Captain's Hill where he was killed by Indians.

Thorfinn Karlsefne's expedition I have shown to have sailed along the west coast of Greenland to Disco Island, across to Cape Dyer in Baffin Land, south past Labrador, east of Newfoundland, past Nova Scotia to Cape Cod, around Monomoy Point to Falmouth; going further south Karlsefne sailed past Rhode Island and into Long Island Sound coming to the island of Manhattan, and here his expedition settled until it found a more feasible place called Hop, which from the saga's description fits the location of Fresh Creek by the Patuxent River on Chesapeake Bay. It is likely that Karlsefne's expedition explored much of the territory between Manhattan and Chesapeake Bay, searching for suitable settling ground.

In tracing all these activities, my prime source has been the sagas recounted in the Hauksbok and the Flateyarbok. By careful interpretation of the sagas with regard to geographical and solar observations, points of sailing, and references to time-distances and natural resources, and taking into consideration the ships of the Norsemen and their knowledge and possibilities of sailing at the time, I have concluded that the legends reveal logical and reasonable accounts of the Vineland voyages which tally remarkably well with landmarks on the North American east coast.

Since the basis for my arguments is the sagas, perhaps it is wise to dispel at the outset any doubts surrounding their validity. I shall therefore briefly give my reasons for believing these doubts to be unfounded.

The Hauksbok was written about 1330–1334, and the Flateyarbok about 1387–1395. Many scholars have believed these books to be the written versions of sagas handed down through generations by word of mouth, and therefore inaccurate and mixed with myth and fantasy. I am in favor of the view, however, that these narratives have been transcribed from an original great saga, 'Saga Eirik's Rauda', now lost, which was written at the time of the voyages, or shortly thereafter, to record and collect the information, perhaps for succeeding voyages. The details of Bjarne's navigation (his ship's journal), for example, could hardly have been composed 400 years after his voyage; and handed down by word of mouth they certainly would not have retained their explicit and tallying features. Bjarne's voyage, furthermore, is given so exactly that not even one of his sailors could have told the story. Only Bjarne, the navigator, the master of the ship, or his mate, could have remembered all these details.

The details of the other voyages also coincide so completely with possibility, and often the only reasonable possibility, that it is hard to believe the records were handed down orally over the centuries still intact. Obvious fiction, for example, in the Flateyarbok story of Thorvald's voyage about a cry and a supernatural warning to leave the land after killing Indians should be recognized as fiction and not be allowed to invalidate the whole story. Such fiction is typical of all narratives of the time and tends to indicate further the genuineness of the story; as I have pointed out the cry could actually have come from one of Thorvald's sentinels, and to clothe it in mystery was the storyteller's prerogative.

Evaluation of the Literature

If my theories are accepted, what about all the other theories that have been advanced; are they to be rejected or are they compatible? If we are to evaluate them we must briefly run through the material that has been collected thus far.

The first histories

In the year 1622, an Icelandic Priest by the name of Tormod Torfaeus was sent from Iceland to the royal library in Copenhagen with some old handwritten documents.[1] For King Frederick IV Tormod Torfaeus then began the task of writing the history of Denmark and Norway in Latin, the scientific language of that time. It was natural for him to write the history of the old Norwegian Island Empire first as he was an Icelander and also because he had brought with him the handwritten documents which he himself could interpret so well. His first work on the Faeroe Islands was published in 1695, and in 1697 his history of the Orkney Islands appeared. His book on Vineland appeared in 1705, and another on Greenland in 1706. These last two books were for a long time the best sources of information about the earliest history of America, and they have been used up to the present time.

In accordance with the international law of the time that the sovereignty of a country belongs to the discoverer, Torfaeus suggested that the Danish-Norwegian King lay claim to North America by virtue of the fact that it was discovered by Norsemen from Greenland. But by 1705 it was too late to make such claims since England, Holland and France were already strongly engaged in the colonization of America.

In 1777, Gerhard Schöning published Snorre Sturlason's *Heimskringla*; to it he added the account of the Vineland travels given in

[1] A. W. Brögger: *Vinlandsferdene*, Oslo 1937, p. 10.

the Flateyarbok, thereby bringing to light more information about the voyages to Vineland.

Schöning's third volume was published posthumously by his friend, P. Suhm, in Copenhagen in 1781. Already at this time Schöning had tried to determine the exact location of Leiv's *Budir* in America through the sun observations taken by Leiv Eiriksson in Vineland. With the aid of Thomas Bugge, Professor of Astronomy, he calculated, by using Páll Vidalin's explanation of 'eyktarstad' meaning a meal taken at 4:30 p. m., that the shortest day of the year at Leiv's *Budir* must have had nine hours between sunrise and sunset, and that, on this basis, without taking into consideration the refractions, Leiv's *Budir* must have been at 41 degrees, 22 minutes latitude. This is the approximate latitude of Falmouth, Massachusetts, and closer computation with respect to the location of Leiv's *Budir* cannot be made, even with modern nautical instruments. (Cf. my discussion of Leiv's sun observation in *Early American History*.)

But critics from 1781 to the present day have discarded Schöning's computations. Until 1830 nothing further of importance was written until Danish authorities decided to publish all the written sources pertaining to the history of Greenland. This challenging task was given to the Icelanders Finn Magnusson, Sveinbjörn Egilsson and C. C. Rafn. Together they published between 1837 and 1845 the great work, *Grönlands Historiske Mindesmærker*, in three volumes.

Grönlands Historiske Mindesmærker is a collection of all that was known about the Vineland voyages since they took place. In order to clarify certain questions about the Vineland expedition, C. C. Rafn sought the aid of American scholars as early as 1829. And in 1837 he published his great work, *Antiquitates Americanae*, in Icelandic, English, French and Latin. Here Rafn makes the first attempt to locate on the American coast all the names mentioned in the sagas about Vineland with an understanding and precision that are remarkable. No other author has had such a broad background and knowledge as Rafn, and the importance of his work should not be underestimated.

With Finn Magnusson he studied all the sources thoroughly, assuming them always to be correct until proven otherwise. This healthy attitude on Rafn's part was the main cause of Gustav Storm's criticism. Storm maintained that Rafn did not distinguish clearly between historical fact and legend. But on the contrary, Rafn's treatment of the old sources, as a historian and as a philologist was unimpeachable; his weakness was that he was no sailor.

In the study of the Vineland expeditions a thorough knowledge of sailing is as important as knowledge of philology. In order to explain Leiv Eiriksson's sailing route from Greenland to Vineland, Rafn drew, as scholars after him have drawn, on a map, a line from Julianehaab on Greenland to Newfoundland. From there on, the line is drawn to Nova Scotia and Cape Cod. Since wild grape vines grew on Cape Cod, Rafn and many other scholars assumed or guessed that this must be Vineland. But sceptical critics have said that this is nothing but guesswork and far from sufficient proof. The same objection has been raised against all other scholars who have tried to fix the exact locations of the early Viking settlements in North America.

In the third volume of *Grönlands Historiske Mindesmærker* pages 885–86, Rafn summarized the results of his research concerning the location of the three lands mentioned in the sagas. 'Helluland it Mikla', he asserts, is the coastal stretch north of the Hudson Strait and the coast of Labrador; 'Littla Helluland' is Newfoundland; 'Markland' is Nova Scotia; 'Vineland' is Massachusetts and Rhode Island; 'Kjalarnes' is Cape Cod. Here we notice that Rafn introduces two Hellulands, the Great Helluland and the Little Helluland. But in the sagas there is mention of only one Helluland which I have located as Baffin Land.[2] Rafn, too, seems to have realized the saga's mention of a single Helluland but it was important for him to extend it as far south as Newfoundland which he established as the third land from which Bjarne Herjulfsson sailed with a southwesterly wind. It seemed obvious to Rafn that this wind took Bjarne from Newfoundland to Greenland in four days, a distance of 148 metric miles or 1,480 kilometers. He found the possible average speed of the Viking ship to be 270 kilometers per day; and by assuming that Bjarne sailed under a full gale, justifying the addition of an extra 100 kilometers a day to the average speed of Bjarne's ship, Rafn found it possible for Bjarne to reach Herjolvsnes on Greenland from Newfoundland in four days.

One should not assume, however, as Rafn did, that because the wind blew in one direction the ship would sail in that same direction.

[2] *Early American History: Norsemen before Columbus*, Oslo 1964, pp. 29, 53 ff., 91 ff.

Another salient point, as I have shown previously, is that Bjarne is much more likely to have crossed over to Greenland from Baffin Land than from Newfoundland. No one at that time would have set course from Newfoundland to Greenland. Except for Kjalarnes (Cape Cod), the location of the other names that Rafn suggests are, in my opinion, incorrect. But Rafn deserves much credit for his work which we will refer to later in this discussion.

Another opinion – Gustav Storm

The first major dissent to Rafn's *Antiquitates Americanae* came with Professor Gustav Storm's *Studies on the Vineland Voyages*, 1884–89. Gustav Storm was a well-known philologist who had translated several Icelandic sagas and had done much fine research. While reading about the Vineland voyages in the Hauksbok he found on one page the heading, 'Saga Eirik's Rauda'. To Storm this was very significant.

Earlier scholars such as Rafn and Magnusson had always regarded the Vineland saga in the Flateyarbok as indicative of a much larger saga of Eirik the Red called 'Saga Eirik's Rauda', believed to be long lost; Gunlaug, the Icelandic monk, had referred to this 'Saga Eirik's Rauda' and also mentioned Leiv Eiriksson's discovery of America when he wrote his own *Kristni-Saga* of the Christianizing of Iceland before he died in 1218. It is Gunlaug's *Kristni-Saga* which was one of the works copied into the Hauksbok by its writer.

Now Storm interpreted his finding in the Hauksbok to mean that here was the real 'Saga Eirik's Rauda', that the saga of Eirik the Red had not been lost, but, on the contrary, had always been the one in the Hauksbok. It is the Hauksbok that attributes the discovery of America to Leiv Eiriksson while the story of Bjarne's discovery is told in the Flateyarbok. Storm thus proposed that the Hauksbok saga, which had previously been named 'Thorfinn Karlsefne's Saga' by Rafn, should be rightfully called 'Eirik the Red's Saga', and the Flateyarbok Saga, named by Rafn 'Eirik the Red's Saga', should be called 'Grænlendingaþatter'.

Many scholars have supported Storm's view here, but I doubt very much that it is reasonable to infer so much from the words, 'Saga Eirik's Rauda' found on one page. Hauk, the writer of Hauksbok, copied indiscriminately all he chose to include, and there may be any number of reasons why the words occur there or why Hauk might have placed them there, perhaps as a reference, or a musing of the author, not necessarily purporting to indicate the Hauksbok saga by name. Gunlaug's mention together of the saga of Eirik the Red and Leiv's discovery of America is not sufficient evidence either that Eirik the Red's saga tells the story of Leiv's discovery and therefore must be the saga in Hauksbok. The fact that Gunlaug does refer to 'Saga Eirik's Rauda' seems to indicate rather that this prior saga did exist.

To continue, however, Storm found the Hauksbok to have been written long before the Flateyarbok and the more authentic of the two. Comparing the two records point by point, Storm concluded that the same incidents must be the subject of both, but that the Flateyarbok was filled with improbabilities. Therefore he suggested that when it was written, 'the main parts of the tradition were already obscured, and that the author had eked out the tradition with his own invention'. Therefore, he continued, the geographical descriptions and statements which occur only in the Flateyarbok must 'be accepted with great care, and only be used where they can be adapted to Eirik the Red's Saga in the Hauksbok'.

Storm acknowledged the discovery of America by the Greenlanders, but accepted as proof the fact that the oldest Icelandic accounts were in accordance with the reference to Vineland made by the Archbishop of Hamburg's chronicler, Adam of Bremen, around 1070, which is the first known authentic reference to the American continent. He placed no faith in the sagas alone and regarded some of them as fictitious.

In support of his theories Storm wrote his *Studies*. Beginning with an investigation of Leiv Eiriksson's sun observations in Leiv's *Budir* given in the Flateyarbok, he thought that the time for 'Eykt', the evening meal, is different for the various places, and for that reason Vineland (meaning Leiv's *Budir*) cannot be located. He concluded that 'The attempt, therefore, from astronomical computations to locate Vineland in Rhode Island, is without any support whatever; and hence the geographical position of the country must be sought in some other way. The first step will obviously consist in examining the original sources of information, to which the following chapters are devoted'.

Professor Storm is incorrect here, however, for he and his astronomer, Geelmuyden, erred in identifying 'Eyktarstad'. In their calculations they regarded 'Utsuders et' (the south-west airt from the old Icelandic law of Grágás) as only the horizon from south-southwest to west-southwest, only half of what it should be (the full quarter of the horizon from south to west), and therefore did not find the right place for 'Eyktarstad'. According to their findings Leiv's *Budir* would have been situated in the middle of Newfoundland on 49° north.³

In *Early American History* I have given an analysis of Leiv Eiriksson's sun observation, based on reference to Snorre Sturlason's Calendar and the calculations of researchers. This shows the location of Leiv's *Budir* to have been at a latitude of Falmouth or Waquoit Bay. We have also mentioned earlier that Schöning and Bugge, already in 1781, had come to this conclusion.

I do not think there is reason to doubt the seriousness of Leiv's sun observation. Leiv Eiriksson was an explorer. It was his and the Greenlanders' intention to explore the three lands which Bjarne had seen, and Leiv had sought advice and support for his expedition from the King of Norway. It would be advantageous and certainly expected that valuable information regarding the latitude (the time of sunset) of the new land be brought back. King Olav Tryggvason, who had traveled all over Europe, would then, himself, be able to make latitude comparisons with England and France. So it is reasonable to assume that the observation of the sun was one of the matters that was planned and studied during the year which Leiv spent with King Olav.⁴ We know also from the sagas that Leif had agreed to proclaim Christianity in Greenland for King Olav, possibly in return for support for his expedition.

It is also likely that Leiv, as leader of the expedition, had to know all that was known about navigation at that time, including observation of the sun. We would expect Leiv to have made the observations correctly, or that the King would have provided someone who could. Perhaps this is one of the reasons why Leiv stopped in Falmouth where the sun sets over the ocean. In any case it would be certainly unfair to consider Leiv an ignorant farmer boy as some have contended.

Storm emphatically rejected the Flateyarbok narrative of Bjarne's voyage. He assumed that Bjarne sailed to Earl Eirik in Norway in

³ See Almar Naess: *Hvor lå Vinland*, Oslo 1954, pp. 55–67. ⁴ *Grönlands Historiske Mindesmærker*, Vol. I–III, Copenhagen 1837–45, vol. II, pp. 223–25.

the year 1001 and found it unlikely that Bjarne would delay visiting the Earl for fifteen years and thereby delay the investigation of the new lands if he really had found them. However, Gerhard Schöning had found earlier that Bjarne sailed to Earl Eirik in Norway in 988–89, only two years after his discovery of the three lands. That Bjarne received no support for the immediate investigation of the new lands may be because Earl Eirik was occupied at the time with defending his interests in Norway (cf. Snorre Sturlason: *Heimskringla*, the battle of Hjorungavag, p. 163).

Storm also maintained that the details of Bjarne's voyage must be fictitious as true details could not be handed down for four hundred years until 1387–95 when the Flateyarbok was written. I maintain, however, that these details, so exact and plausible, could not have been handed down but must have been recorded, still fresh, in the lost 'Saga Eirik's Rauda' from which the Flateyarbok was copied.

Professor Storm was most incredulous of the story of the grapes collected in Vineland, especially as he thought Vineland was Nova Scotia where no grapes grew. He felt such fantasy was completely invalidating and reason enough for rejection of the whole saga in the Flateyarbok. He then altered the text in the saga in order to demonstrate how ridiculous it was to believe that the Greenlanders could have found grapes in Vineland: 'They gathered the grapes in the springtime in such quantities that they filled the boat with grapes, and the vines they spoke of as big trees which they cut to use as lumber.'

There is no doubt from the saga (cf. Gathorne-Hardy's translation of Leiv Eiriksson's discovery of America in *Early American History*) that Leiv and his crew gathered grapes in the autumn, as soon as they had begun to investigate the land around the camp, and not in the springtime as Storm retold the tale. And why should they cut down vines? Would Storm have us believe that the Norsemen were so ignorant of vines that they would think vines could be used as lumber? The wild grapes grew, no doubt as they grow today, up along the trees in the woods, sometimes the vines growing over the tops of the highest trees. For that reason they had to cut the vines, and sometimes the big trees as well in order to reach the grapes. On this point too, the saga does not seem to be amiss. Storm tells us furthermore that Tyrker's joy at discovering the grapes was because he became drunk eating them, this to demonstrate again the saga's fantasy in assuming a Turk to have been on the expedition. Tyrker the Turk could well have been on the expedition as I have indicated in my earlier book *Lysstreif over noregsveldets historie*.[5]

After rejecting the Flateyarbok as a source, Storm began to locate Helluland, Markland and Vineland on the American coast using the Hauksbok as his only reference, and a dictionary definition of a *doegr*'s sailing as twelve hours.[6]

But this, too, is erroneous for the Norsemen regarded a *doegr*'s sailing on the ocean to be a full day, approximately twenty-four hours.[7] Thus Storm was short of twelve hours' sailing distance per *doegr* when he reconstructed Thorfinn Karlsefne's journey. He plotted the journey from Fiskarnesset in Greenland, skipping Disco and Baffin Land (Helluland) and suggested that Karlsefne laid a southerly course to Labrador. In this way Storm lost track of Karlsefne's expedition and could only guess the route. He suggested that Helluland was Labrador, that Markland was Newfoundland, and that Vineland was Nova Scotia.

Storm could not believe that the Norsemen had sailed farther south than Nova Scotia. Thus he located Kjalarnes at Cape Breton on the north side of Nova Scotia; Furdustrandir on the east coast of Cape Breton Island, and Straumsfjord as the Strait of Canso or a nearby inlet. Vineland (Leiv's *Budir*) he placed in southern Nova Scotia. My earlier reconstruction of the Karlsefne voyage would show Storm's guesses to be quite incorrect.

Storm clearly accepted Leiv Eiriksson's discovery of Vineland and Thorfinn Karlsefne's colonization of the land. But he could not believe that they had sailed so far south as Cape Cod or Rhode Island and thought it a great mistake to believe the story of the grapes and the fruitfulness of the land. In his desire to preserve the reliability of the sagas Storm cut out all he thought was fiction and considered only what he could be sure was true. But Storm took sailing details too little into consideration and so misjudged the possible extent of the Vineland voyages.

[5] Norges Svalbard- og Ishavs- Undersökelser, Meddelelser no. 56, Oslo 1944, pp.132–34. [6] Storm explained in a note on page 337 of his *Studies* that according to Fritzner I², 282, *doegr* signifies day or night in Old Norse. [7] J. Kr. Tornöe: *Early American History*, op. cit., pp. 17; A. Naess: op. cit., pp. 177–78.

Grassland, not Vineland?

We continue our brief survey of the history of history and come in 1910 upon Sven Söderberg, philologist.

Upon discovering that 'vin' in old Norse could mean 'meadow' or 'grassland', Söderberg interpreted 'vin' in Vinland as meaning 'meadow' or 'grassland', not 'wine' or 'grapes'. In Norway there are many such names involving this first definition: Hundvin, Langvin, Marvin, Skodvin. Accordingly Söderberg contended that Vinland meant Grassland.

It was his opinion that Leiv Eiriksson had not found grapes in Vineland. And he dismissed as mistakes the saga writer's mention of Leiv or Tyrker finding grapes in Vineland, and Adam of Bremen's reference to grapes in Vineland.

Several writers, who for various reasons would like to locate Vineland in Hudson Bay, on Labrador, the St. Lawrence River, Newfoundland or Nova Scotia, adopted more or less this idea of Söderberg's. Nansen's use of the idea in order to show the sagas were only fairy-tales lent it the authority to persist for the next fifty years. Today, even such well-informed writers as Helge Ingstad still consider the idea.[8]

More recent investigations, however, suggest that it is a misunderstanding to take 'vin' as meaning meadow or grassland. Accordingly O. Heitmann Andersen, in his *Det norske folks Busetning og Landnåm belyst ved Stedsnavnene*, argues that originally in Gothic 'vin' was a descriptive term that meant 'beautiful' or 'nice', but that long before Leiv Eiriksson's time the word had dropped out of the language, so that even if it had come to be used specifically for grasslands and meadows, there was no chance of Leiv Eiriksson using it in this way (pp. 99–137). Furthermore, the Icelandic scholar Ólafur Lárusson, in his *Island, Nordisk Kultur*, V, (1939) (p. 64) remarks that names beginning or ending with 'vin', 'tveit', 'setur', 'vangur', or 'rud' are not to be found in Iceland.

In northern mists

In 1911 Fridtjof Nansen published his famous work *In Northern Mists*, wherein he devoted a chapter to the Vineland voyages. Like Gustav Storm he believed that the Greenlanders discovered America,[9] but at the same time he tried to prove the sagas about

---

[8] Helge Ingstad: *Landet under Leidarstjernen*, Oslo 1959.   [9] In an address to the

Vineland were only folklore and myth, and that Vineland was a legendary land.[10]

Some scholars have taken Nansen's words very seriously and many have been confused by them: 'Even if certain of these countries are legendary, as will presently be shown, it must be regarded as a fact that, in any case, the Greenlanders and Icelanders reached some of them, which lay on the northeastern coast of America, besides Greenland, about five hundred years before Cabot (and Columbus)' (*In Northern Mists* vol. I, p. 312).

Nansen's efforts to prove the sagas legendary were based on several prime factors: distrust of the oldest historical (Adam of Bremen's) reference to Vineland, lack of more substantiating evidence regarding Vineland and its discovery in the oldest Icelandic authorities, the writing of the sagas so long after the events are said to have happened, and the mythical and romantic air of the sagas and their likeness to existing myths of the time.

If Nansen is to dispose of Adam of Bremen's reference to Vineland, he must dispose of much supporting evidence of its reliability. Adam's very brief reference alone may not carry much weight, but seen in the light of the time, it takes on the air of testimony. Adam of Bremen was a conscientious historian who wrote down faithfully all he was told in his *History of the Archbishopric of Hamburg*. The King of Denmark had himself told Adam that 'there was an island in that ocean visited by many, which is called Wineland, for the reason that vines grow wild there, which yield the best of wine. Moreover unsown grain grows there in profusion, and this we know is not a fabulous fancy, for the accounts of the Danes prove that it is a fact'.[11]

The King of Denmark who told this story was Svein Estridson,

Royal Scottish Geographical Society, Nov. 9, 1911, and published in *The Scottish Geographical Magazine* vol. 27, Nansen stated his opinion regarding the Norsemen: '. . . They found their way to the White Sea, and the lands beyond; they discovered the wide Arctic Ocean and its lands; they settled in the Scottish Islands; found and colonized the Faroes, Iceland, Greenland; were the discoveres of the Atlantic Ocean and of North America . . .'. [10] We do not know to what extent, but it is likely that Nansen was influenced by his assistant, Professor of Folklore, Moltke Moe. In a Foreword to a research work written in Copenhagen by Björnbo and Petersen these relate that Nansen also accepted *their* proposal to check over his manuscript.

[11] Adam of Bremen; cf. Einar Haugen: *Voyages to Vinland, the first American Saga*, New York 1942, pp. 97–100.

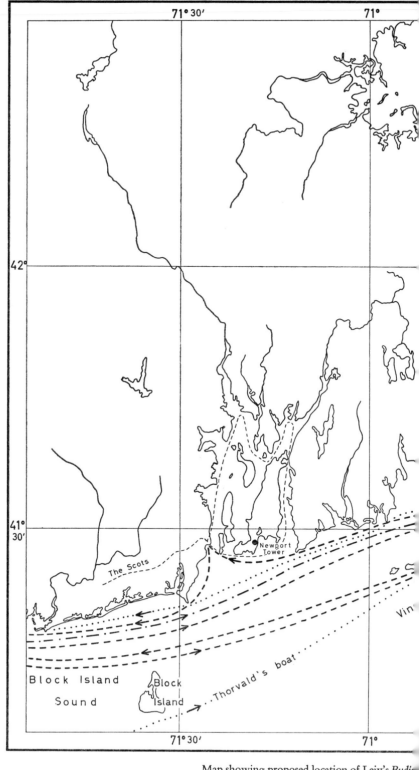

Map showing proposed location of Leiv's *Budi*

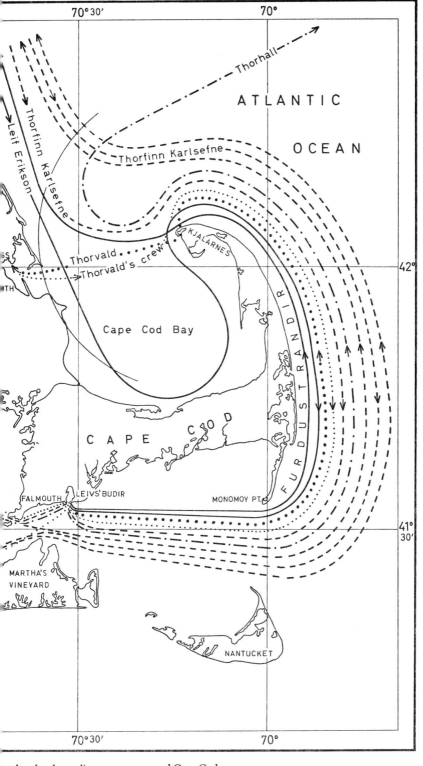

Routes taken by the earliest voyagers round Cape Cod.

and we know of him that he was a traveler and an observer, wise and well versed in letters, and known to Adam as a man who 'retained in his memory all the deeds of the barbarians exactly as if they had been written'. From such a reliable source Adam would take down the reference even though he may have had doubts about its truth. We know that, earlier, men from Iceland had been on the American coast and that they brought their stories back to Iceland from where the stories spread to Norway and Denmark.

The discovery of Vineland in all its details was recorded in the sagas. Should these sagas not be given some consideration as authority? For those who wrote alongside or after the Vineland sagas would consider these stories already told and only copy them or refer to them as Gunlaug did in his *Kristni-Saga*. And if we disregard the Vineland sagas, of course we have not much evidence left at all.

We have already discussed and supported the theory that the present accounts we have of the Vineland voyages in the Hauksbok and the Flateyarbok were copied from earlier writings and therefore the dates of the two cannot be considered the dates when the sagas were first 'made up'.

We can see that, regardless of the mythical fabric into which the details of the Vineland voyages may have been woven, these details of sailing, course, speed of ships, geography, when picked out coincide so completely with what we know to be true, that we must give proper consideration to the sagas.

The fact that the characteristic mythical features of wine and wheat appear in the Vineland sagas may be an interesting comparison for folklorists, but it does not automatically place the voyages in the category of myths. The discoverers had called the land Vineland because they found grapes there. This was the simple and unobtrusive mention of grapes; the accounts of the discoverers do not indicate any intent to symbolize a land of goodness by use of the metaphorical 'grapes'; and it is both unfair and unreasonable that we should attribute to them this intent and thus consider the tales fables.

The attempt to place Vineland among the legendary Insulae Fortunatae, the happy lands of the dead, is indeed in bewildering contradiction to the acknowledgement of the fact that the Norsemen did land on Vineland.

Is it so impossible to believe the Norsemen traveled the North

American coast, reaching the fertile area around Falmouth and Martha's Vineyard? Is the only other explanation that the story is a fable? Certainly it is no fable that Nansen accomplished the quite impossible feats of skiing across Greenland, and across the drift ice toward the North Pole, reaching as far as 86° 14' and then Franz Josef Land!

Objection

William H. Babcock was one of the first to object to Nansen's work, in his book *Early Norse Visits to North America*, published by the Smithsonian Institute, 1913. Babcock contributed many interesting details in his opposition to Nansen, but mainly followed Gustav Storm's view of the sagas and the location of Helluland and Markland. On the location of Vineland he sided with Rafn and placed Leiv's *Budir* in Mt. Hope Bay, Rhode Island.

Professor William Hovgaard published, in 1914 (New York), his book *The Voyages of the Norsemen to America*. He felt that some of the sources in the Flateyarbok could be useful in determining the location of the various names in the sagas. Tending to follow Gustav Storm's view he thought that Helluland was Northern Labrador, Markland was Southern Labrador and Vineland was Newfoundland.

In Copenhagen in 1917, Professor H. P. Steensbye published *The Norsemen's Route from Greenland to Vineland*. Like Gustav Storm he rejected all sources in the Flateyarbok and hoped instead to locate the names in Thorfinn Karlsefne's Saga from the Hauksbok. Steensbye found Helluland to be Northern Labrador, and Markland Southern Labrador to the Strait of Belle Isle. From there he sent Karlsefne's expedition into the St. Lawrence River, suggesting Newfoundland as Bjarney, Hare Island as Straumsey and the southern side of St. Thomas on the St. Lawrence as Hop.

I have shown earlier that Karlsefne's expedition never went into the Gulf of St. Lawrence, but sailed south along the east coast of Newfoundland (cf. *Early American History*).

The saga says Karlsefne traveled four *doegr* from Disco to Markland changing course from south to southeast from Helluland to Markland. We have shown this change to have occurred between Baffin Land and Labrador where the coast lies southeast. Steensbye thought the change in course to be significant and would have it occur at the Strait of Belle Isle allowing Karlsefne to sail into the

St. Lawrence River. Changing from south to southeast here, however, would lead into the Atlantic, and so Steensbye thoughtfully altered the saga's text so that Karlsefne changed from south to *southwest*, sailing right into the St. Lawrence. But after such devices, Steensbye's further interpretations can have little interest for us, and the same can be said of his followers, Fossum, Hermannsson and Thordarson.

G. M. Gathorne-Hardy

We come now to a milestone in the literature on the Vineland voyages, G. M. Gathorne-Hardy's *The Norse Discoverers of America. The Vineland Sagas translated and discussed*, 1921.

Gustav Storm and Nansen had criticized and interpreted the Vineland sagas in a manner that seemed to leave little truth in them. Gathorne-Hardy started out to prove that there was no material in the sagas that could not have derived from first-hand acquaintance with the American coast, and his book from pages 113 to 195 has a number of cogent arguments to this effect.

As a conclusion (page 245) he says: 'Whatever criticisms have been passed upon Bjarne's voyage by those who are unable to bring it into line with their theories, this voyage would be regarded, as in itself, sufficient to substantiate the fact of Norse discovery'.

After a successful explanation of the sagas' texts, Gathorne-Hardy started to reconstruct Bjarne's route from Iceland to America and finally to Greenland. Gathorne-Hardy got the idea that he could plot Bjarne's route backwards from Greenland to the point where he first came into the American coast. The saga gives the information that Bjarne sailed with a southwest wind from the third land to Greenland. So Gathorne-Hardy substituted the direction of the wind for the course of the ship. In that way he drew a line on the map from Greenland and reached Newfoundland. In a stricture upon Professor Hovgaard's book, *Voyages of the Norsemen to America*, Gathorne-Hardy says (p. 239): 'The whole point of giving the direction of the wind (southwest) is to supply an indication of the course. To this course Professor Hovgaard pays no attention'.

To this remark we can add the information that Hovgaard was a naval officer before he became a professor, so he knew that he could not substitute the direction of the wind for the course of the ship. We quote Gathorne-Hardy further (p. 246): 'The simplest way of dealing with this voyage [Bjarne's] is to plot it backward from Greenland... Taking the data arrived at in chapter V for the length of a "dœgr sigling", we may plot the distance represented by the unit at about 150 miles. The wind, we are told, was southwest. Plot from Herjolvsnes (Sermisok) in the south of Greenland four "Dœgr" units in a southwesterly direction and then draw a land-form which will serve for the "island" which was the third land seen, follow its coast to a point five more "dœgr" units southwest. Lastly mark land on the course at the end of the five days and also two days from the end. The result will be as shown on the shaded portion of [my] sketch. These indications are quite near enough to the truth to show pretty conclusively that the "lands" were the Barnestable peninsula (Massachusetts), Nova Scotia and Newfoundland, respectively'.

As we have seen my reconstruction of Bjarne's journey in *Early American History* outlines it quite otherwise, locating the three lands as Nova Scotia, Newfoundland-Labrador, and Baffin Land respectively.

Is it reasonable to think Bjarne could have tried to sail from as far south as Cape Cod or Newfoundland to Greenland? When he approached land after drifting in the Atlantic, he did not know how far west he had come as he had no map or compass and he had never been in these waters before; nor could he have known where Greenland was located with respect to this new land. All he knew was that he was too far south, as the terrain of his first land indicated. So all he could do was head northward until he came to a latitude similar to that of Greenland. He could not set a course straight to Greenland not knowing where Greenland was, and he would not take a chance in hoping to find it by sailing out to sea. And though he might feel he knew in which direction he wanted to go, as a sailor he would know the difficulties of keeping on his course. For a sailing ship always deviates to some extent from its course because of tides, currents and its own leeway, even with a good compass. Bjarne had no compass and when the weather was cloudy he would have no sun either; and he would not know anything about his course. This was exactly how he had got lost in the first place. If by chance he did reach Greenland, he would not have known how to find the new lands again. The only reasonable solution for Bjarne was to sail north, and coming to Baffin Land where there were glaciers, he would be more assured of finding Greenland by sailing eastward on this latitude.

Let us suppose with Gathorne-Hardy that Bjarne first approached the coast of Cape Cod. He would have to have, or wait for, clear weather to know anything about the direction of the coast and also get his bearing to the north from the sun. But in clear weather from Cape Cod, Bjarne would discover the coast of Massachusetts. From Cape Cod he would sail north along the coast to Nova Scotia, as he could do that without a compass, even in cloudy weather. Thus, Nova Scotia could not be Bjarne's second land, as he could see the continuous coast all the way to Nova Scotia. On a northerly course from Cape Cod, Bjarne would reach the coast east of Portland, and on a more easterly course he would reach the coast of Maine or the Bay of Fundy. There is no reason to doubt that he could sail from Cape Cod to Nova Scotia without losing sight of land.

Accordingly, if Cape Cod to Nova Scotia was one land, Newfoundland could not have been the third land but the second; and as even Gathorne-Hardy considers Newfoundland-Labrador to be one land, the third land must be Baffin Land.

Gathorne-Hardy suggests that Bjarne laid a northeasterly course from Cape Cod because he had a southwesterly wind. But nothing is mentioned in the saga about the wind or the course when Bjarne left his *first* land; it was when he left his second land that the wind was southwest. It is essential here to remember that Bjarne had no means of determining that his course to Greenland should be northeast; he could only be sure he was too far south and that he must lay his course northward.

The saga states Bjarne sailed in five days from his first land to his third land. If Baffin Land was his third land, he could not have reached it from Cape Cod in five days. But from Nova Scotia he could well sail to Baffin Land in five days. And it is also northern Nova Scotia where we have 'the island north of the land' where Leiv Eiriksson landed when he came to Bjarne's first land. There is no island north of Cape Cod.

I think the discussion so far has been sufficient to indicate Gathorne-Hardy's reconstruction of Bjarne's voyage to be incorrect. We may conclude here, as I have done earlier, that no one sailed from Newfoundland to Greenland, but that the Norsemen sailed from Baffin Land to Greenland when they returned from Vineland; from Greenland to Vineland, they sailed the same course in reverse. To navigate directly from Greenland to Newfoundland was too difficult without a compass, a log, or a clock. It was safer to cross over to Baffin Land and sail down the coast with two minor jumps over the Hudson Strait and the Cabot Strait. For these two stretches of water the ship could wait in harbor for favorable wind and weather.

It has been the points of navigation that have been misunderstood by most of the scholars in this field, and which therefore have led them astray.

Gathorne-Hardy attempted also to clarify the route of Thorfinn Karlsefne. He discovered that this route did not coincide with his suggestion that Bjarne and Leiv crossed back and forth between Greenland and Newfoundland. Thus he suggested a possible error in the Karlsefne saga, a mistake in transcription of the direction southeast for southwest, the latter being more consistent with what he thought was the course of Karlsefne's journey.

On page 232 he says:

'But setting the question of local resemblance apart, the identification is defended on the ground that one text gives for the direction from Helluland, "they changed their course from south to southeast". This seems to me a most unreliable statement on which to found a definite and positive conclusion. In the first place, the change of course indicated is only given by Hauk; the purer companion version states merely that the explorers had a north wind. Having regard to the fact that the word "southeast" (landsuðr) occurs in the very next sentence – "an island lay to the southeast" – there is here an obvious trap for the unwary copyist. Supposing the word in the archetype of the saga to have been originally southwest (utsuðr), a course more consistent with the general direction of Karlsefne's investigations, it is extremely likely to have been mistranscribed with a word so like it close at hand to catch the eye. Besides, the courses on the whole are so manifestly wrong, or at best vague approximations, that no one can be on sure ground who relies on them.'

This presumption Gathorne-Hardy made in order to support his mistaken belief that Newfoundland was Helluland. If Karlsefne laid a southeasterly course from Newfoundland, he would go into the Atlantic Ocean and that does not sound correct. But Karlsefne was not in the latitude of Newfoundland when he changed his course; he was sailing across the Hudson Strait from Baffin Land to Labrador. (Cf. *Early American History*.)

But Gathorne-Hardy was not the first to suggest the crossing from

Newfoundland to Greenland. The first was C. C. Rafn in *Antiquitates Americanæ* and *Grönlands Historiske Mindesmærker*. We may call this Rafn's school, and his followers are Packard, Kohl, Horsford, Gathorne-Hardy, Hjalmar R. Holand and Frederik J. Pohl. They adhere to this theory because it is the shortest distance between Greenland and Newfoundland, and they find it the only way that the Greenlanders could reach as far south as Cape Cod. But just the same the theory is surely wrong.

Other opinions

M. M. Mjelde in his article 'Eyktarstadproblemet og Vinlandsreisene' published in *Historisk Tidsskrift*, 1927, tried to work out a new theory for finding 'Eyktarstad'. He suggested that it should be on the horizon 60 degrees west of south.

Almar Naess tried to prove Mjelde correct and calculated that Leiv's *Budir* could not have been further north than 37 degrees on the American coast.[12] This means that Leiv had his camp somewhere in Chesapeake Bay. There is no doubt that Leiv could sail as far south as Chesapeake Bay along the American coast, but the description in the saga of the place where Leiv built his houses fits only Falmouth as I have shown. Mjelde's 'Eyktarstad' differs too much from Snorre Sturlason's 'Eyktarstad' which coincides exactly with Leiv's 'Eyktarstad' in Falmouth.[13] For these reasons we reject Mjelde's theory.

V. Tanner in his *De Gamla Nordbornas Helluland, Markland och Vinland* (1941) suggests that Labrador could be Vineland, discounting the grapes in the saga.

In 1944 I published a book dealing with the long-distance voyages of the Greenlanders and Icelanders, *Lysstreif over Noregsveldets Historie* (Oslo). It was here that I first proposed my location of Bjarne's lands: Vineland as the coastal country from the Gulf of St. Lawrence toward the south, Markland as the coastal country from the Gulf of St. Lawrence to Hudson Strait, and Helluland as the present Baffin Land. Subsequently I decided to go to America to investigate and locate the rest of the places mentioned in the sagas, which research led to *Early American History* and this book.

A host of other writers have contributed more or less to the litera-

[12] A. Naess: op. cit., pp. 81–101. [13] J. Kr. Tornöe: *Early American History*, op. cit., pp. 65–70.

ture of the Vineland voyages. They have mostly been followers of one or another of the men we have just discussed and on the whole contribute little that is new, with some exceptions:

Hjalmar R. Holand has undertaken a great amount of historical research mostly in connection with the Kensington Stone; A.W. Brögger has listed chronologically the literature from 1436 to 1936 on the Vineland voyages, and has reviewed all of the most significant; John R. Swanton and Johannes Bröndsted both concluded that there is virtually no agreement among scholars as to the locations of the different place names in the sagas.

We must also mention Helge Ingstad's book, *Landet under Leidarstjernen*, in which he locates Leiv's *Budir* in northern Newfoundland. According to my investigations Leiv's *Budir* is situated in East Falmouth. Recently, however, Mr. and Mrs. Ingstad excavated the foundations of eight houses in L'Anse aux Meadows. This must be a camp built by hunters from Greenland or belonging to Landa-Rolf's colony in the years 1289-95. At the time of going to press we are awaiting Mr. Ingstad's report on what in any event must be considered a most important find. Gwyn Jones in his *The Norse Atlantic Saga* (Oxford 1964) also suggests Newfoundland as Vineland, which certainly cannot be correct.

The bibliography at the end of this volume includes the authors discussed and indicated here as well as the works referred to throughout the book.

When surveying the literature we can see that there is little agreement among the many authors on the subject. This is largely due to the fact that most writers have failed to use the saga as a guide in locating the different names mentioned. Most of them have picked a more or less arbitrary point on the map where they believed that the expedition had passed, – e.g. Cape Cod for Bjarne's voyage, which Bjarne never actually reached – thus disrupting the very foundation for their theories from the start.

Many authors just took a ruler and found the shortest distance to be a line between Newfoundland and Greenland – and then attempted to make this line fit the saga. This is indeed wrong, because the Norsemen would not have sailed by the shortest route here as it was too difficult and hazardous. They sailed along the coast northward to Baffin Land, a somewhat longe route, but by far the safest.

Some scholars have even changed the saga's text where it did not fit in with their construction of the route. They have suggested how errors must have entered when the saga was told by word of mouth, or when it was transcribed.

Others have ignored certain facts in the sagas when these did not fit in, for example, with their location of Leiv's *Budir*. This was begun by Sven Söderberg in *Snællposten* in 1910 and has continued up to this year in the accounts by Gwyn Jones and even Helge Ingstad, when he tries to fit his recent find in Newfoundland to the saga's description of Leiv's *Budir*; in place of the grapes mentioned in the saga both these present-day writers bring in meadows or grasslands up north in Newfoundland or Labrador.

But surely it is now time to end this guessing game and to take the sagas at their word.

A Leiv Eiriksson Memorial

There is a matter for which this conclusion has some relevance and of which I may add a brief mention, and that is the late Mr. Winters Haydock's proposal to erect a memorial to Leiv Eiriksson where the Norsemen landed. On hearing of my own investigations Mr. Haydock was eager to press forward with this project. Since I believe that the historical support for this project and its location in Falmouth has been established, there is now only need for active support. To that end I conclude by quoting from the correspondence that took place between Mr. Haydock, Mr. Soyland, Editor of *Nordisk Tidende*, and myself.

> From Winters Haydock
> Member American Society of Civil Engineers.
> Member American Institute of Planners.
>
> February 22. 1957
>
> Dear Mr. Soyland:
>
> I am writing you at the suggestion of Dr. Raymond Dennett, President of the American-Scandinavian Foundation.
>
> I am proposing the introduction in Congress of legislation to provide for the establishment of a national park and museum in honor of Leif Ericson. Its location, as I now visualize the legislation, would be determined by a Commission of Scholars, to be appointed by the President, on the basis of the opinion of the majority of the Commission as to the *most probable* location of Leif's colony. This plan recognizes the controversial aspect of the problem and the possibility that a positive answer, to which all scholars would agree, can never

be found; and if we waited for unanimity of opinion we would probably never get the park and museum.

In my opinion, as things stand today, the most probable location is that suggested by Mr. Frederick Pohl in his book 'The Lost Colony' which is Fallin Pond at the head of Bass River on Cape Cod. I have no fixed ideas on the subject, but Pohl's argument seems somewhat more convincing than that, for example, of E. F. Gray in his book 'Leif Ericson, Discoverer of America', published in 1930. Furthermore, the Bass River area lends itself well to the ultimate development of an ambitious plan for a spectacular national memorial, accessible to millions of our citizens. As I have indicated, I do not feel that we should wait for uncontrovertible proof of the correctness of this or any other location, proof which probably will never be forthcoming.

One aspect of my plan, which would probably be more than selfsupporting, would be an amphitheater in which would be held throughout each summer a pageant depicting the events of the discovery and of contemporary Viking life, on somewhat the same general order as the pageant written by Paul Green and produced every summer on Roanoke Island, N. C. to commemorate the 'lost colony' founded there in the latter part of the sixteenth century.

I will greatly appreciate any suggestion on this subject you may care to offer.

Sincerely yours,
Winters Haydock

Mr. Soyland passed this letter on to me and I wrote to Mr. Haydock expressing my interest in his plan, but pointing out my belief that Leiv's *Budir* was at Falmouth, and not Fallin Pond or Bass River. Mr. Haydock replied:

June 12. 1957.

Dear Mr. Tornoe:

Your letter of June 8. concerning my proposal for a national park as a memorial to Leif Ericson was of great interest to me. I had been disappointed in not receiving a reply to my letter of February 22. on this subject to Mr. Carl Soyland and had been forced to conclude that he was not interested in the matter. I had written him at the suggestion of Mr. Raymond Dennett, President of the American-Scandinavian Foundation who stated that Mr. Soyland's group was interested in revitalizing the Leif Ericson Memorial Committee. Mr. Dennet thought that Mr. Soyland's group would be the place to look for the greatest support for the plan . . .

Because of the condition of my health I am no longer able to take the lead in this matter . . . I would like to find someone to push my ideas. Although I know nothing about you other than information conveyed in your letter, it has occurred to me that perhaps you could take on the job. I would think that

Mr. Dennet would be willing to help. I think that any Scandinavian group should be interested.

Briefly, my idea is that a bill should be introduced in Congress to provide for the appointment by the President of a commission of qualified persons who would be charged with the duty of suggesting the most appropriate location for a national Park in honor of the landing on American soil of Leif Ericson.

Since it is possible that we will never have positive proof as to the location of his landing, which would be agreed to by all scholars, the opinion of a mere majority as to the *most probable* location should be required. Otherwise we might never get action. The law should include a time limit for the Commission's report.

Either in the same or in later legislation provision should be made for a park of ample size, to include a museum and also, perhaps, accommodation for an institute concerned with Scandinavian and other European cultures contemporary with the period of the discovery and concerned with all evidence of pre-Columbian landings on American soil by white men. As you know, there is evidence that landings may have been made by other Europeans, including Irish and Welsh.

The park should also include an outdoor amphitheater for an annual pageant depicting the events of the landing. (This should include a replica of the type of ship used by the Vikings for ocean travel.) The pageant should be of somewhat the same order as that written by Paul Green of the University of North Carolina and produced every summer in Manteo on Roanoke Island, N. C., to commemorate the first English colony in America, which so mysteriously disappeared. This pageant, called 'The Lost Colony', lasts all summer and is entirely selfsupporting.

I regret that your book on early American history has not been published yet. I assume that it contains the evidence as to the Chesapeake Bay landing. I had understood that such evidence exists.

Since I am no longer able to push my proposal for a national park, perhaps you can take it up.

Sincerely yours,
Winters Haydock

I hope that this plan will be realized.

Chronological Survey Regarding Geographical Discoveries Made from Iceland and Greenland

Year
982 Eirik the Red sails from Iceland in a northwesterly direction to Blåserk in Greenland and from there to Julianehaab in western Greenland. (Cf. J. Kr. Tornöe: *Lysstreif over Noregsveldets Historie.*)
985 Eirik the Red sails the same route back to Iceland. (Cf. op. cit.)
986 Eirik the Red sets out with 25 ships to colonize Greenland; founds Brattahlid at Julianehaab. (Cf. op. cit.)
986 Bjarne Herjulfsson tries to sail the same route to Greenland but gets lost and sails past Nova Scotia, Newfoundland, Labrador and Baffin Land before reaching Greenland. (Cf. J. Kr. Tornöe: *Early American History: Norsemen before Columbus.*)
998 The Icelander, Thorgils Errabeinsfostre, is shipwrecked far north on the East Greenland coast. (Cf. *Floamanna Saga.*)
999 Thorgils is trapped in the field ice all summer. (Cf. op. cit.)
1000 Thorgils travels southward to Seleyar (probably Angmagssalik, eastern Greenland). (Cf. op. cit.)
1001 Thorgils continues to Rolf (probably Lindenows Fjord). (Cf. op. cit.)
1001 Leiv Eiriksson sails to Vineland, establishes Leiv's *Budir* at Falmouth, Mass. (Cf. *Early American History.*)
1002 Leiv Eiriksson returns from Falmouth, Mass. to Greenland. (Cf. op. cit.)
1002 Thorgils reaches Eirik the Red in Greenland (Julianehaab). (Cf. *Floamanna Saga.*)
1003 Thorvald Eiriksson sails to Leiv's *Budir*. (Cf. *Early American History.*)
1004 Thorvald dispatches the boat expedition with orders to sail southward around Vineland and over to the west coast. The

voyage lasts five to six months including the return journey to Falmouth. (Cf. op. cit.)

1004 Thorvald is killed by the Indians in Plymouth, Mass. (Cf. op. cit.)

1005 Thorvald's crew returns from Falmouth to Greenland. (Cf. op. cit.)

1006 Thorstein Eiriksson's attempt to sail to Vineland is unsuccessful. (Cf. Flateyarbok.)

1007 Gudrid, Thorstein Eiriksson's widow, comes back to Brattahlid to live with Eirik and Leiv. (Cf. Hauksbok.)

1007 Thorfinn Karlsefne comes to Brattahlid. (Cf. op. cit.)

1008 Thorfinn Karlsefne marries Gudrid in Brattahlid and is chosen as leader for the expedition to Vineland. (Cf. op. cit.)

1009 Thorfinn sails to Vineland, stops at Straumsey (Manhattan Island), sails to Hop (Patuxent River). (Cf. *Early American History*.)

1010 Thorfinn trades and battles with the Indians in Hop. (Cf. op. cit.)

1011 Thorfinn returns from Hop to Straumsey (Manhattan). (Cf. op. cit.)

1012 Thorfinn decides to reverse his voyage back to Greenland. (Cf. op. cit.)

1012 The brothers Helge and Finnboge come from Norway to Greenland. (Cf. Flateyarbok.)

1013 Helge and Finnboge join Freydis, Leiv Eiriksson's sister, and sail to Vineland. (Cf. op. cit.)

1014 Freydis returns from Falmouth, Mass. to Greenland. (Cf. op. cit.)

1019 The highest official of Greenland, Skald-Helge sails from Herjolvsnes to Greipar (the fjords north of Scoresby Sound) in order to punish some robbers. (Cf. *Grönlands Historiske Mindesmærker* vol. II p. 440.)

1025 Pope John XIX orders Archbishop Unwan of Hamburg to organize the church in Greenland and the adjacent islands (Helluland, Markland and Vineland). (Cf. op. cit.)

1027 Lika-Lodin usually went hunting in the summer along the coast of Greenland. Sometimes he found dead sailors and brought their bodies to the church in the settlement, hence his nickname Lika-Lodin (Corpse-Lodin). King Olav Harald-

	son orders Lodin to bring the bodies of Finn Fegin and his crew from Finnsbudir back to Norway. (Cf. op. cit.)
1047	The Arctic expedition of King Harold mentioned by Adam of Bremen.
1066	Lika-Lodin meets King Harold in Norway. (Cf. *G. H. M.* vol. II.)
1121	Bishop Eirik Upse sails from Greenland to Vineland. (Cf. Icelandic Annals 1121.)
1129	Sigurd Njålsson hunting along the coast of Greenland finds the body of Arnbjörn and his men at Hvitserk and brings their bones to the church in Gardar. (Cf. *G. H. M.* vol. II.)
1160	A papal letter to the Archbishop of Trondheim concerns the people of an island twelve days and more from Norway and other lands. (Cf. Part I, Through the Catholic Church.)
1189	The Greenlander, Asmund Kastanrazi, comes to Iceland from Krosseyjar and Finnsbudir after hunting in the summer. (Cf. *Lysstreif...*; Icelandic Annals 1189.)
1194	The discovery of Svalbard (Spitsbergen). (Cf. *Lysstreif...*; Icelandic Annals 1194.)
1200	The shipwreck of the priest Ingemund Thorgeirsson on the East Greenland coast. (Cf. *Gudmund Aresöns Saga*; Icelandic Annals 1200.)
1250	*The King's Mirror* is written including discoveries made by the Greenlanders and Icelanders.
1261	King Haakon, after agreement with the Greenlanders, extends the limit of his kingdom theoretically to the North Pole. (Cf. *Lysstreif...*)
1266	The expedition of the priests from Gardar to the Kane Basin. (Cf. *Lysstreif...*)
1285	The two priests Adalbrand and Thorvald discover Newfoundland (and Duneyar). (Cf. Icelandic Annals 1285.)
1289	King Erik of Norway sends Rolf (Landa-Rolf) to Iceland to announce the King's wish for people to join Rolf in colonizing Newfoundland and perhaps Duneyar. (Cf. Icelandic Annals 1289; *G. H. M.* vol. III.)
1290	Landa-Rolf organizes, with the backing of the King, an expedition for colonizing Newfoundland and Duneyar. (Cf. *G. H. M.* vol. III p. 12.)
1295	Landa-Rolf dies in the land he has colonized. (Cf. op. cit.)

PART II

SUBSEQUENT VOYAGES WITH REFERENCE TO COLUMBUS

Columbus in the Arctic?

When I first started this work on the early American history I confined it to the Norse settlements, not planning to include Columbus. But in 1959 when I had almost finished my manuscript about the Norse discoveries, a Russian historian revealed that he had found in Russian archives a letter from Columbus. In this letter Columbus rejected the rumor that he was seeking China when he set out on his famous journey from Spain in 1492. He contended that he knew about the islands in the Caribbean Sea before he started from Spain, and he understood that it would be of great value for Spain to possess these islands.

On Monday, October 12, 1959, the *New York Herald Tribune* printed the following article on page 3:

Soviet Historian Declares Columbus Tricked World

A Soviet Historian said today that Christopher Columbus hoodwinked the world 467 years ago because he knew all along where America was.

The historian, identified only as Tyspernik, a lecturer at the Kazakh Pedagogic Institute, was quoted by the Moscow radio as saying he had discovered a 'secret letter' from King Ferdinand and Queen Isabella of Spain to Columbus in which they decided to 'whitewash reality' and pretend that Columbus had discovered a new world of riches.

Actually, Tyspernik said, Columbus already knew the location of the Antilles, where he made his landfall on October 11, 1492, and merely dressed up his 'discovery' story to make colonization more attractive to Spaniards.

Tyspernik said other sailors had been to America and had told Columbus all about it, and that when he found only a wild land which did not 'make a favourable impression' on the crews of his ships, the Nina, Pinta and Santa Maria, he and the Spanish government decided to circulate the 'version' that 'fabulous wealth' had been discovered.

Tyspernik said he believes Columbus 'misled historians of his day into believing that he was trying to find a new route to the Orient, when all the time he knew he was headed for America'. He explained away Columbus's diaries detailing the voyage by saying, 'on his return (to Spain) from the voyage, he deliberately altered the contents of his diaries, and the most difficult thing for me was to find out what he had written in them before'.

Simultaneously the following report appeared in *Aftenposten*, Oslo:

Did Columbus know about America before his journey in 1492?

Moscow Broadcasting station yesterday contended that a Russian historian found a letter from Christopher Columbus which changes the theory that the discovery of the Antilles was a surprise to Columbus. The historian, who is a member of the academy of Uzbek, maintains that Columbus knew very well about the West Indies before he started on his famous journey in 1492. The historian has found a letter from Columbus to Queen Isabella which shows that Columbus knew the location of the Antilles and understood the value they would be to Spain in the future.

The story that Columbus had set out to discover a new route to India was made up by the Spanish court in order to explain the reason for the great expenses to outfit the expedition.

On the 5th of September 1960, John Wingate reported, in his radio broadcast, several statements from a congress of 1,500 historians which met in Russia to deal, *inter alia*, with Columbus's discovery of the Antilles. It was revealed that the Columbus letters affirmed that Columbus had learned the position of the Antilles from a Portuguese.

It was precisely these reports that led me to investigate the status of the Greenland colony at the time of Columbus. The next step was to try to consider the contact between Greenland and Vineland. Another important question was if there could be any possible contact between the Norsemen and Columbus. We have the note from Columbus (see p. 73) where he contends that he had been on the 73rd parallel north. What other evidence there is and what new light this brings to bear on Columbus I hope will become apparent to the reader.

Greenland's Bond with the New World

What became of the new lands of North America whose discoveries were recounted in the sagas? Were they only discovered and preserved through the sagas, or were they traveled or colonized to the extent that contact with them was sustained and knowledge of them remained?

It has been customary to say that the Greenland settlements had disappeared and, through them, knowledge of the Vineland voyages forgotten by the time of Columbus. We shall try to trace the history of the Greenlanders and their activities in North America from where the Vineland sagas cease to tell us more – about the year 1015 – and the Greenlanders are newly converted to the Christian religion.

Through the Catholic Church

From *Grönlands Historiske Mindesmærker* we know that in 1025, Pope John XIX ordered Archbishop Unwan of Hamburg to organize the Church in Greenland and 'the Adjacent Islands'. 'The Adjacent Islands' must be Helluland, Markland and Vineland, supposed to be islands at the time; unfortunately we have no record of activities, however. In 1112 we know from Icelandic Annals that the Pope created Eirik Upse Bishop of Greenland and the Adjacent Islands and Eirik traveled to Vineland in 1121. It is safe to assume a Bishop would not go to Vineland unless there were a significant number of Norsemen there. Bishop Eirik did not return from his journey, hence we still have no reports of life in Vineland.

But we do have a letter from about 1160 from the Pope to the Archbishop of Trondheim in which the Pope gives dispensation from Canon Law regarding marriage for the people of an island (*insula quedam*) situated twelve days sailing or more from Norway (*a Norwegia*), and likewise from all other Christian lands (*ab aliis terris*). These people were too closely related to intermarry and could not legally marry Indians or Eskimos as they were heathens and of different race considered inferior to the Nordic race and called 'skraelings' by the Norsemen. The nearest land where the islanders could find legally marriageable Christian women was situated twelve days' sailing or more from the island.

Where is this island then; could it be Vineland? One of the first to attempt to solve this question was Knut Robberstad who thought the island probably had to be Greenland (*Frå gamal og ny rett*, Oslo 1950, pp. 35–36). Later Odd Nordland concluded that the island probably was Greenland (*Viking*, Oslo 1953, pp. 87–107). Walter Holtzmann, Arne Odd Johnsen and Sigurd Grieg also wrote about this island and came to this same conclusion. (Cf. *Historisk Tidsskrift* no. 4, 1961, pp. 136–38, pp. 160–74, Einar Molland.)

In 1959 Eirik Vandvik published a book, *Latinske dokument til norske historie fram til år 1204*, and in it set forth a transcription of the letter from the Pope and his own hypothesis that the island could not be Greenland but could be a settlement of Greenlanders in Markland.

From the transcription of the Pope's letter:

> Alexander III Trundensi archiepiscopo.
>
> Ex diligenti relatione nunciorum tuorum nostris est auribus intimatum, quod insula quedam a Norwegia per xii dietas et amplius distats posita esse et metropolico iure tue ditioni subesse perhibetur, cuius itaque parrochiani ita se consanguinitate uel affinitate dicuntur contingere, quod matrimonium uix secundum statuta canonum legitime possint contrahere ... Verumtamen si ab aliis insulam prescriptam terris per xii dietas sicut audiuimus constet distare et eius populo tantam in hiis necessitatem noueris iminere, tu ascitis tibi suffraganeis tuis cum suo et aliorum religiosorum uirorum consilio poteris dispensare et supradicto populo, ut in v⁰ et vi et vii gradu contrahant matrimonium ...

Rendered in translation the letter is as follows:

> Through the news considerately conveyed by your messengers, it has been made known to us that there lies an island twelve days' journey and more from Norway, and which is subject to your metropolitan authority. The parishioners there on the island are said to be so near in consanguinity and affinity that they can hardly contract marriage legally in accordance with the statutes of the Canon Law ... But if it is as we hear that the above mentioned island lies twelve days' journey from other lands, and you know for certain that the islanders live in such great difficulties with regard to this, then you may inform your

bishop and with the advice of yourself and other religious men give dispensation and allow the aforementioned people to contract marriage in the fifth, sixth and seventh degree...

Vandvik concluded (p. 171) that the island must be so small or so little known that the Curia did not find it necessary to mention it by name, but described it only by its distance, twelve days' sailing or more from Norway and all other Christian lands. Among the Norwegian settlements there is no island which could be said to lie twelve days' journey from both Norway and other lands. Greenland does not fit the description for the journey from there to Iceland is much shorter than twelve days and much shorter than the journey to Norway. However, if the Norse Church province of Greenland is interpreted to mean Norway, the island could be a settlement of Greenlanders in America, preferably in Markland which was thought to be an island.

Bishop Eirik Upse had gone to Vineland in 1121, perhaps to a colony of Christian settlers from Greenland or to convert skraelings in America (cf. allusion to the newly converted in the letter). A Norse colony on the American coast would have belonged to the bishopric of Gardar in Greenland and so was under the metropolitan of Norway, the Archbishop of Trondheim. Such a colony would have had great difficulties because of the Canon laws of marriage, and the nearest Christian land would have to be Greenland. From Labrador to Gardar the voyage could have taken about twelve 'doegr' (= halvdöger – day or night).

Vegard Skånland, working from a second version of the Papal letter (both versions must be transcriptions of the original assumed to have been destroyed during the Reformation) concluded that since it was explicitly stated in the letter that the island concerned had only one bishopric (*episcopatus . . . subiectus* – used in the singular), and at the same time lay twelve or more days' journey from Norway, it is without doubt clear that the letter must refer to Greenland.[1]

Second transcription of the Pope's letter:

Alexander tertius Trundensi archiepiscopo.

Ex diligenti relatione nuntiorum tuorum nostris est auribus intimatum quod insula quedam a Norwegia per xii dietas et amplius

---

[1] Vegard Skånland: 'Supplerende og kritiske bemerkninger til Eirik Vandvik: Latinske dokument til norsk historie fram til år 1204'. *Historisk Tidsskrift* no. 4, 1961, pp. 137–38.

distat, in qua episcopatus tibi metropolico iure subiectus existere perhibeatur. Cuius itaque parrochiani ita sibi inuicem dicuntur contingere quod matrimonia inter se uix aut nunquam legitime possunt contrahere, presertim cum difficile sit eis – et pauperibus fere impossibile – terram hac occasione exire et aliunde uxores quererre. Quare a nobis iamdicti nuntii tui instantius postulant ut illius insule populo debeamus super hiis utilius prouidere et aliquam inde facere dispensationem.

In translation:

Through the news considerately conveyed by your messengers it has been made known to us that there lies an island twelve days' journey and more from Norway, and which is subject to the metropolitan jurisdiction of your bishop, whose parishioners therefore are said to be so related to one another that they can scarcely or never contract legal marriages, especially since it is difficult for them – and for the poor almost impossible – to leave the land for this purpose and seek wives elsewhere. Wherefore your announcements just mentioned demand that we should provide for the people on that island and make some dispensation there.

From the Pope's letter

I propose that a third location, Martha's Vineyard, must be the island referred to by the Pope. I believe Vandvik's location lies closer to the truth but placing it in Markland does not coincide with the information that the bishop Eirik Upse in 1121 journeyed to Vineland not to Markland. Surely the bishop would travel to a colonized place, which seems to be Vineland, the land from the Cabot Strait southward as far as the Norsemen knew it at the time.

The Papal letter's reference to a single bishop on the island does not necessarily restrict the island to Greenland as Skånland concludes. If bishop Eirik Upse lived for some time in Vineland at Martha's Vineyard as I propose, then Martha's Vineyard was an island with one bishop during the time of Eirik's stay.

If, as Vandvik says, the Norwegian Church colony on Greenland is regarded as Norway – and here it should be added that legally Greenland was a part of the realm of Norway[2] – then Martha's Vineyard fits the location of the island for it is at least twelve days' journey from Greenland and at the same time at least twelve days' journey from all other Christian lands.

[2] J. Kr. Tornöe: *Lysstreif over noregsveldets historie*, Norges Svalbard- og Ishavs-Undersökelser, Meddelelser no. 56, Oslo 1944, pp. 199–206, p. 215 of the English summary.

In *Early American History: Norsemen before Columbus* we have traced Bjarne Herjulfsson's voyage from Nova Scotia to Greenland in nine days. The distance from Martha's Vineyard to northern Nova Scotia (Scatari) is roughly 600 miles; at the approximate speed of 8 knots or a little more, the speed we usually find in the sailing directions at that time over several distances, the additional time it would take to sail from Martha's Vineyard to Scatari would be approximately three days, making the entire journey from Martha's Vineyard to Greenland approximately twelve days. The evidence seems to be quite indicative.

We are told further of the mass emigration from Greenland to America in 1342 and that Greenlanders left the Christian faith and went over to the heathens of America.[3] The expedition of Paul Knutsen (1355–63) was sent to America by the King of Norway to search for these emigrants in order to reconvert them to Christianity. It is possible that the expedition went first to Martha's Vineyard and from there north to Hudson Bay searching for emigrants traveling south by the Nelson River. Tracing the route of Knutsen's expedition has been difficult and to this end we refer to Hjalmar R. Holand's book: *A Pre-Columbian Crusade to America*, New York 1962. In this book (p. 43) Dr. Roberto Almagio is quoted as saying: 'An expedition of this nature, the purpose of which, furthermore, is not apparent, if carried out in the manner supposed by Holand, could be conceived of only if it had been preceded and prepared by a long series of more limited but repeated reconnaissances towards the interior, concerning which there is not the slightest evidence'.

I shall show later that all the information Paul Knutsen needed was known to the Greenlanders. They had been working the woods in the Hudson Bay region for 200–300 years and they certainly would have explored the Nelson River far inland; it is very likely that they traveled over the Nelson River far into the heart of America and therefore the Paul Knutsen expedition chose to go that way. Evidence of the Greenlanders' knowledge of the Hudson Bay region exists also in their reference to Hudson Bay as Markland's Botnar; *Botnar*, signifying bottom, indicates they had investigated the area well enough to know there was no westward opening from Hudson Bay, see note 33, page 64.

[3] *Grönlands Historiske Mindesmærker* vol. I–III, Copenhagen 1837–45, vol. III, pp. 459–64.

Holand's research into the Kensington Stone has met with a good deal of scepticism and it has been suggested that the Stone was a hoax carved by the farmer who claimed to have found it or the local schoolteacher. It is my opinion, however, that the Stone very likely is genuine. Mr. Holand has, I gather, considered only the Paul Knutsen expedition as responsible for the Stone's inscription; this may be only one explanation. The Greenlanders were a relatively large population in need of wood, charcoal, pine tar and so forth (cf. pp. 52 ff.), and therefore perhaps sent several expeditions a year to the Hudson Bay and Nelson River regions. It is likely that these expeditions traded with the Indians and Eskimos (cf. note 10, p. 51). It is possible also that some German merchants or an English priest or monk may have accompanied some of these expeditions and perhaps one of them may have drafted the text of the Stone.[4]

Holand has also pointed out that the Newport Tower is an old fortified Norse Church[5] which lends evidence of Norsemen living in Rhode Island, or nearby, for example, Martha's Vineyard. Though no substantiating information is known yet, Nantucket Island, being farther away from the mainland, and thereby from the Indians, than Martha's Vineyard, might have been another likely place for settlement.

In any case it is fairly certain that traffic from Greenland to America continued in the fourteenth century and that Greenlanders emigrated in order to settle in America.

Newfoundland is an island

There is further evidence in the discovery of Newfoundland as an island in itself. Earlier we have seen that all expeditions to Vineland sailed east of Newfoundland which they thought was connected with Labrador, observing the Strait of Belle Isle to be a fjord. If they had sailed through the Strait of Belle Isle they would have had to give it a name and there would have been four lands instead of three in the west. So it remained until 1285, when Icelandic Annals reveal that the two priests Adalbrand and Thorvald discovered Nyaland.

This means that they must have sailed through the Strait of Belle Isle and around Newfoundland and seen that it was an island.[6]

[4] J. Kr. Tornöe: *Lysstreif* . . ., pp. 185–191. [5] Hjalmar R. Holand: *America 1355–1364*, New York 1946. [6] C. C. Rafn: *Antiquitates Americanae*, Copenhagen 1837 and 1841, pp. 259–63, 451–52.

The Norse name was Nyaland or Nyfundnaland which C. C. Rafn suggests Cabot later anglicized to Newfoundland. The entry in Icelandic Annals is supported by a younger note in Are Frode's *Landnámabók:* 'Hafa vitrir men sagt at suðvestr skal sigla til Nyaland (Nyfundnaland) undan Krysuvikur bergi' [Well-informed men have said that south-west is the course from Krysuvikur Bergi (little east of Reykjanes, Iceland) to Nyaland (Nyfundnaland)]. This course from Iceland to Newfoundland is correct.

The distance from Reykjanes to Cape Race, Newfoundland is about 1,440 miles. Is it possible that the Norsemen could sail straight from Iceland to Newfoundland at this time? It could be done if they had the 'Leidarstein' (compass). The compass was known in Europe by 1195 and surely it could be known to the Scandinavians about a hundred years later. But did they have ships big and strong enough then to make the journey?

According to recent excavations on Bryggen in Bergen there existed in Bergen, about 1250, some ships which were larger and much stronger than the Gokstad ship of earlier Vikings. Referring to the report we find that 'according to the preserved beam the length of the ship may be estimated at about 90 feet, the width 24 feet'.[7] With such ships and a compass for navigation it should have been possible to sail from Iceland straight to Newfoundland.

Colonization

The German historian, Gebhardi, mentioned in 1778[8] the discovery of Nyaland: 'A remarkableness in the life time of King Erik is the discovery of a new shore line in northern America, which an Icelander, Rolf, was the first to discover in 1258 and he lived there from 1289 to 1295 when he died'. Though we do not know Gebhardi's sources, he has, however, related the discovery to North America; and we know that nowhere else could a new land have been found after 1250, and that Newfoundland was, until sailed around, still undiscovered as an island.

Gustav Storm suggested that Landa-Rolf set out to colonize a spot on the east coast of Greenland. This certainly cannot be correct for the Greenlanders and Icelanders knew there was no place along that

[7] Asbjörn E. Herteig: 'The excavation of 'Bryggen', the old Hanseatic Wharf in Bergen', *Medieval Archaelogy*, vol. III, p. 185.   [8] *Kongeriget Norges historie* vol. I–II, Odense 1777–78.

coast suitable for colonization; they had been traveling back and forth along the east coast of Greenland for the past 300 years. Neither would they have named it a new land, nor is it likely that people who had heard of Vineland would settle on a barren coast of Greenland.

Gebhardi has stated that Landa-Rolf discovered Duneyar and Nyaland in 1258. Perhaps therefore Rolf received the King's support to colonize the land. But Icelandic Annals have 1285 as the year of discovery by Adalbrand and Thorvald; could it be that Gebhardi has interchanged the ciphers 5 and 8. Both authorities agree, however, that Landa-Rolf, with the backing of the King, started to colonize the land in 1288 and he lived in Nyaland until he died in 1295. Where could this colony have been located?

For climatic and agricultural reasons it would most likely be on the south coast of Newfoundland on the peninsula between Fortune Bay and Placentia Bay where there were also several islands abundant in seals, walrus, whale and birds. The island Duneyar is mentioned together with Nyaland and could probably be Miquelon Island. Perhaps an investigation in this area might reveal relics of old Norse farms from about 1300.[9]

We can well understand that it was difficult and dangerous to sail from Iceland to Newfoundland in 1300. The colony was small and had few products to sell, and the rewards did not warrant the risk. In order to encourage men to make the journey and to make that journey more worthwhile, the Pope granted indulgences to seamen who sailed to the islands on the other side of the frozen ocean: 'Et primo indulgentiis Navigantibus ultra mare glaciale ad insulas concedendis'. 'The islands' refers to Greenland, Helluland, Markland, Nyfundnaland and Vineland, and there must have been a colony significant enough in the west for the Pope to want to maintain contact with it.

[9] The waters of northern Newfoundland were good hunting grounds for walrus and seal which passed through the Strait of Belle Isle. In the summer months hunting parties from the south may well have traveled to the north and made camp there. The recent discovery of traces of old Norse settlements by Helge Ingstad at L'Anse aux Meadows in northern Newfoundland is especially interesting in this connection. It is possible, in my opinion, that investigations along the coast of Labrador may reveal further indications of Norse settlements of this type and also perhaps of settlements of earlier Greenland hunters and woodsmen prior to Landa-Rolf's colonization.

Some old bills of lading describing various hides brought from Greenland to Europe were collected by the Archbishop of Trondheim. The bills of lading contained names of such animals as elk, beaver, black bear, ermine, glutton, lynx, otter, sable and wolf, the hides of which the Greenlanders could only have obtained by hunting in Newfoundland or in the Canadian areas, or by trading with the native Indians, since most of these mammals never existed in Greenland.[10]

We must certainly agree, then, that by the fourteenth century the Greenlanders and Icelanders had not lost interest in or contact with the lands of North America. They had maintained their hold to the extent of colonizing the land, and records were available in Iceland at the time though unfortunately some of the records were later destroyed.

[10] Herluf Winge: *Grönlands Pattedyr*, Meddelelser om Grönland vol. 21, Copenhagen 1902, p. 322.

The State of Greenland

The opinion that knowledge of America did not persist but receded into legend until it was all but forgotten for practical purposes exists on the basis of what I consider the mistaken belief that the Greenland colony – the link to America – was very small and subject to severe hardship, and had gradually died out by the fifteenth century and the dawning of the 'age of exploration'. On the contrary from what we know of the Greenland colony it existed into the sixteenth century and became a significant source of information for the Kings of Europe.

Population

Most of the books on Greenland picture a small population of about two to three thousand, sitting quietly in Greenland waiting for starvation. Nothing can be more false. We find that the population in Greenland at 1250 was one third of a bishopric elsewhere according to the Archbishop of Norway, Einar Gunnarson, in *The King's Mirror* written about 1250.[11] In this year Norway had at least 560,000 people in five bishoprics. The average population of the Norwegian bishopric would be 112,000 and one third of that would be 37,333 which would be the approximate population of Greenland.[12]

Such a population seems reasonable when we learn from Icelandic Annals that Greenland had at least sixteen churches and two monasteries. Wilhjalmur Stefansson[13] says of Iceland: 'It is thought that 50,000 settlers, most of them from Norway or Ireland, moved to Iceland between 870 and 930 . . .'; we know further that Iceland in 1095 had a population of 78,000. Should it not be expected, then, that Greenland would have more than a few thousand settlers at its peak period?

---

[11] There is no reason to doubt the Archbishop's reliability here as he received the tithes, church fines and head taxes from his several bishoprics. [12] J. Kr. Tornöe: *Lysstreif* . . ., p. 199. [13] *Northwest to Fortune*, 1958, p. 12.

No people at the time were so traveled as the Greenlanders. They frequently visited Europe for trade or attended church conventions. Some of them traveled all the way to Rome and even to Jerusalem;[14] Greenlanders certainly participated in the Crusades in the twelfth century and very likely they were represented amongst the 'verings' (the Emperor's bodyguard) in Constantinople.[15] They had traveled so far to the north that they knew that Greenland was an island and that there was no land north of it.

The Greenland colony was not at all limited to Greenland alone. Greenlanders sailed all the way to Spitsbergen to hunt walrus; some of them hunted as far north as Smith Sound and the Kane Basin; others hunted in the Hudson Bay regions and parts of Labrador. In 1261 the Greenlanders secured from King Haakon the agreement that they would have the king's protection even if they traveled as far as the North Pole, in effect, no matter where they went in the north, and Norse laws ruled that territory.[16]

The need for resources

We well understand that the need for wood was great in forestless Greenland with a population between 30 and 40 thousand. Wood was used for houses, 'hjeller' for hanging fish, furniture, shipbuilding, tools, fuel and for the preparation of steel; we know today that steel was produced in Greenland and in order to do this, great quantities

[14] The sagas tell us that Gudrid, the wife of Thorfinn Karlsefne, traveled to Rome and told the Pope of the new land Vineland which they had found; the highest official of Greenland, Skald-Helge, made the trip twice, and later Bjarne of Stokkanes and others traveled to Rome as pilgrims. Björn Einarsson journeyed three times to Rome, sailing the last time from Greenland to Israel.

[15] Snorre Sturlason relates in *Magnussönnenes Saga* (1103–1130) that when the sons of Magnus became kings, there came back from Israel, and some from Constantinople, men who had journeyed on voyages such as Skofte Ogmundsson's through the Strait of Gibraltar and the Mediterranean; they were highly regarded and could tell many things. The Kings then equipped an army for a crusade to Israel with sixty ships under the leadership of King Sigurd. The army sailed in 1108 for London, France and then Galizaland in Spain. In Portugal King Sigurd helped Count Henry regain Cintra from the Moors. It is interesting to note this early joint venture of the Scandinavians and Portuguese in view of the fact that more than four centuries later the two peoples would again unite in enterprise (cf. later discussion beginning p. 58). King Sigurd and his army continued on to Lisbon, battling with the Moors again, and then to Sicily, Israel and Constantinople before returning to Norway in 1111. [16] J. Kr. Tornöe: *Lysstreif...*, pp. 68–76, 128–40.

of wood and charcoal were needed. Pine tar for ropes and for the protection of their ships was needed as well.

The closest timber or lumber area to Greenland was the Labrador and the Hudson Bay region. Icelandic Annals tell us of the Greenland ship that had been to Labrador (Markland) in 1347 for lumber, but had drifted off course in a storm, reaching Iceland instead of returning to Greenland. This ship had a crew of eighteen men which was unusually large since only six to eight men were needed to load and sail the ship. This would indicate that the remaining ten men were added to the crew as lumberjacks or as workers to prepare the timber or charcoal on the spot prior to shipping. It is interesting to note here how the Greenlanders lessened their transportation problems to a great degree. They produced the charcoal and the tar at the immediate area where the wood was felled and this consumed less transportation space than the wood needed to produce the materials in Greenland. Perhaps one day remains will be found indicative of one of the areas where charcoal and tar were made.

Working the new lands

We must assume that the farmers in Greenland, as they came to the Hudson Bay region, claimed the forest or land in much the same way as their forefathers in Norway and Iceland had done before them, according to the old Norse laws. The farmer would find a spot suited as a harbor for his ship, preferably where the wood reached close to the shore for easier loading. He could then claim the area as his, if he was the first to do so, and upon his return to Greenland, he could announce the limits of his stake at the *Thing* (Parliament); the law would then protect the area as his property. The Greenland farmers depended on the Hudson Bay and Labrador areas as a major source of wood as the distance from Greenland to Labrador was considerably less than the distance from Greenland to Norway.

It is even possible that the Greenlanders brought hay from the woods as forage for their cattle just as the farmers in Norway made hay for the winter forage while living at their summer grazing sites. The Greenlanders could not take all their cattle with them to the Canadian woods, but they could take their families, returning at the end of the summer to care for their cattle and live in Greenland during the winter. This would explain very well the report of Ivar

Bårdson who came to the western settlement in 1347 and found cattle but no people. This also coincides with the time (1347) when the Greenland ship drifted over from Labrador to Iceland proving that the Greenlanders did go to Labrador for wood.[17]

The Greenlanders needed so much wood that sailing from Greenland to the Canadian areas had to be undertaken every year; people could travel back and forth almost when they wished. It was possible for trappers and hunters to cut lumber in the autumn, leaving it to dry by the next year, and during the winter hunt and trap different kinds of animals. The man who had claimed some ground and woods also had the right to fishing and hunting on his grounds. He could rent this right to others for a certain sum or for a part of their catch. And this could easily bring them in conflict with the Indians and the Eskimos who certainly would not recognize such a right. The hunters also supplied the settlements with drift wood for different use on the farms.

About 1250 the Norsemen had to reduce their hunting activities in North Greenland because of insufficient drift wood and sent their hunting expeditions more often to Labrador and Hudson Bay where, after hunting, they could complete the cargo with timber.

In 1261 the Norwegian king Hakon Hakonson claimed the sovereignty over these territories all the way to the North Pole. After this the Norsemen could claim hunting grounds and woods according to the Norse law. We have no records of hunting in these woods, but we have evidence to this effect in the bills of lading collected by the Archbishop of Trondheim; and we would naturally assume that the Greenlanders would hunt in the Canadian woods as in any other.

But there is also a further indication of activity on the part of Norsemen in the areas round Hudson Bay. In 1266 (cf. *Lysstreif...*, p. 71) an expedition of priests was sent by the Bishop in Gardar to investigate the arrival in Smith Sound in North Greenland of a number of Eskimos. It is reasonable to suggest that the cause of the Eskimos coming up to Smith Sound in the first place was their being pushed out of their hunting grounds in the Hudson Bay area. When the Norsemen proclaimed the territory (Marklands botnar) as their own the Eskimos automatically became outlaws.

[17] G. H. M. vol. III, p. 14.

Expansion to the new lands

The Canadian woods were also the only possibility for expansion for the Greenland population. We can therefore be quite certain that some of the people went over to the Canadian areas for shorter or longer periods. And I believe evidence of them exists in the steel tools and other implements indicating Norse origin which have been found in the Hudson Bay region and attributed to the Eskimos.[18]

The Italian professor Corrado Gini in *Naturen* (No. 7, Bergen 1957) mentions the unusually cold climate which occurred in the fourteenth century together with a possible attack of Agrotis occulta as a reason for the disappearance of the Greenland population. Professor Fægri seems to agree with this opinion. It is reasonable to assume that these misfortunes would hit the western (northern) settlement first and most severely. Lack of forage and wood in Greenland would force the Greenlanders and their families to the Hudson Bay area or Labrador for these supplies in the short summer season. We know that big ships were being used at least to the end of the fourteenth century (Björn Einarson had sailed from Greenland to Israel in 1389) and some cattle could very well have been transported in such ships to the Canadian areas.

Perhaps we have a better understanding here also of the 1342 report regarding the Greenlanders who left Greenland and went over to the heathens in America. If the Greenlanders found it easier to make a living in the Canadian areas, they probably did not return to Greenland. Such emigration could have gone on for a long time and could account for the whereabouts of the disappearing population of Greenland. I consider it all too probable that over a period of time the Greenlanders spread little by little over such vast areas that they could not fail to lose the battle against nature and the more numerous Indians.

We also refer to the blonde Eskimos discovered by Wilhjalmur Stefansson northwest of Hudson Bay where he lived for a year. Stefansson contended that these blonde Eskimos are mixed descendents of the Norsemen. (See also Henry A. Larsen: *Henry med det store skipet*, Oslo 1964, pp. 24 and 32.)

But yet in the fifteenth century the Greenland settlement was alive and active; Icelandic Annals report still nine churches and nine congregations in the Eastern Settlement of Greenland in 1448. And in this century – the century of European exploration, the century of Columbus – Greenland began to become involved in the affairs of Europe.

[18] J. Kr. Tornöe: *Lysstreif...*, pp. 178–91.

Greenland in the 'Age of Exploration'

Since the year 1261 the King of Norway, by agreement with the Greenlanders, sent at least two ships per year to bring goods back and forth from Greenland. About 1270 King Magnus Lagaböter declared the Arctic Ocean 'mare clausum' in order to protect trade with Greenland. No foreigner was allowed to sail further north than to the Faeroe Islands unless he had the King's permission.[19] In 1425 and 1431 King Erik, who ruled a united Denmark, Norway and Sweden, repeated this resolution.

To protect the Greenland colony, the Danish-Norwegian kings made several treaties with the English kings banning English sailors from Iceland and Greenland. Because of the harm wrought by British pirates in Iceland and Greenland, war broke out between Denmark-Norway and England in 1467.[20] The admirals Pining and Pothorst participated in this war, patrolling the coast of Greenland and Iceland against enemy raiders. As neither charts nor lighthouses existed at this time, it was important to have as pilots Greenlanders and Icelanders who knew the treacherous coastal waters at first hand. *Purchas Pilgrims* III, p. 520 says: 'Item Punnus and Pothorse have inhabited Island certayne yeeres, and some times have gone to sea and have had their trade in Greenland.'

This indicates that for several years Pining and Pothorst had their naval base in Iceland and traded in Greenland as they patrolled the coast there, chasing away ships which had no license to sail these northern waters; and they built a landmark on the rock of Hvitserk[21] to signify that the land was under the sovereignty of the Danish-Norwegian King and that foreign sailors had no right to trade or sail there.

[19] J. Kr. Tornöe: *Lysstreif...*, pp. 139–40, 108-09; Arnold Raestad: *Kongens strömme*, Kristiania 1912. [20] A. W. Brögger: *Vinlandsferdene*, Oslo 1937, p. 147. [21] J. Kr. Tornöe: *Lysstreif...*, pp. 23–39, Hvitserk; Sofus Larsen: *The Discovery of North America*, Copenhagen 1924, p. 32.

Scandinavian-Portuguese collaboration

It happened that King Erik of Scandinavia was married to the cousin of Prince Henry the Navigator of Portugal; and through this relationship developed the joint cooperation between the Crowns of Scandinavia and Portugal to search for a western route to China and India which was of prime interest to the trade of both Scandinavia and Portugal.

In 1424 at the Court of Emperor Sigismund, King Erik met Dom Pedro, the brother of Henry the Navigator who had collected information about a new route to China and India. Erik, who ruled Iceland and Greenland, certainly possessed some knowledge about Helluland, Markland and Vineland; he could order his captains, who sailed to Iceland and Greenland, to collect information from the people and from books about the western lands and the route to these lands; he probably provided Henry the Navigator with the map of the Scandinavian countries made by the Danish geographer Claudius Clavus. Later Dom Pedro visited England and the Netherlands and several authors contend that he also visited King Erik in Copenhagen.[22] We may assume that books, maps and letters concerning the western lands would have been shipped from Copenhagen to Portugal at this time. It would not be difficult to translate into Latin the Vineland sagas from the Flateyarbok and the Hauksbok for Prince Henry's use.

Henry the Navigator's interest in a new route to the east was of major concern to him and accordingly his friendship with the Scandinavians – the rulers of the northern waters – would seem advantageous to the far-sighted Prince who poured over his maps and instruments at his fortified haven of Sagres. Sofus Larsen tells us of the Dane, called by the Portuguese Vallarte, who, probably at Prince Henry's instigation, came in 1448 to Henry's court and was by him appointed leader of an expedition to Cape Verde.[23] In 1458 another Scandinavian named Laaland, sent to the Portuguese court again through the efforts of the navigator Prince, took part in the Portuguese war against the Moors in Africa.[24]

---

[22] Sofus Larsen: op. cit., p. 26. [23] Op. cit., pp. 7–16. [24] Op. cit., pp. 16–21. In connection with Portuguese–Scandinavian efforts along the African coast it might be mentioned that a Scandinavian fortress, Christiansborg, was established on the Gold Coast for the purpose of trade, according to Lektor Otto Ottesen. Though this fortress is known to exist in more recent times, it

A joint expedition

Though Henry the Navigator died in 1460 before realizing his probable plans of investigating the northern waters, his policy was continued, and Alfonso V of Portugal and the Scandinavian King Kristjern I succeeded in dispatching the Pining–Pothorst expedition about 1470–1473. Nothing was known of this joint expedition before Dr. Louis Bobé in 1909 found a letter from Carsten Grip, the Burgomaster in Kiel, to King Kristjern III, which informs us that the Pining–Pothorst expedition was undertaken at the request of the King of Portugal.

Carsten Grip relates '... that the two skippers Pyning and Pothorst, who were sent out with some vessels by your Royal Majesty's Grandfather King Kristjern I at the request of his Royal Majesty of Portugal for the purpose of discovering new islands and lands in the North, have erected and made a large landmark (baa) facing the sea and looking towards Sniefeldsiekel in Iceland on the rock Wydthszerk off Greenland, because of the Greenland pirates...'[25]

Dr. Sofus Larsen became interested in the letter from Carsten Grip regarding the cooperation between the Kings of Portugal and Scandinavia and his investigation revealed that Gaspar Fructuoso (1522–90), who wrote family histories, had this to say:

'As the vice-royship of Terceyra was thus vacant on account of the death of the first vice-roy, Jacome de Bruges, thereupon there landed at Terceyra two noblemen who came from the land of stockfishes [Terra de Baccalao] which they had gone out to discover by order of the Portuguese King. One called himself Joao Vaz Cortereal and the other Alvaro Martins Homem'.[26]

The request of the King of Portugal here confirms that referred to by Carsten Grip. Carsten Grip must have had some knowledge about a joint Portuguese-Scandinavian expedition to America about 1470, and he seems to be a dependable man, being not only burgomaster but commisioned by the King to buy maps and books; he must have been known as a well-informed man with certain historical and geographical knowledge.

would be interesting to discover its origins and whether it can be traced back to the time of Henry the Navigator. [25] My translation from the German in Louis Bobé: Aktstykker til Oplysning om Grönlands Besejling, *Danske Magazin* ser. 5, Vol. VI. pp. 303-11. [26] Sofus Larsen: op. cit., p. 70.

A Portuguese views the New World

The question arises: Could Joao Vaz Corte-Real have gone with the Pining-Pothorst expedition by order of the King of Portugal? Sofus Larsen found the name of Joao Vaz Corte-Real on some old maps placed as the name of a bay and a peninsula indicating he had been around Labrador (which area is known for its abundance of stockfishes, i. e. cod). He found also that Joao must have been on his journey made at the King's request about the years just prior to 1473 and he was known to be back in Portugal in the autumn of 1473. The Portuguese King had neither requested nor sent any expedition in that period other than that of Pining-Pothorst. All evidence points to one and the same journey, the expedition at the request of the King of Portugal and undertaken by the Scandinavians. This is our first evidence also that a Portuguese had been in the northern waters and had seen the new lands, and of course he would take this information back with him to Portugal.

Joao Vaz was the father of the three brothers Corte-Real who became very famous explorers; it is likely that their persistence to reach the North American lands is accounted for by the knowledge that their father had seen and been to these lands.

Greenland is still a link to America

The Pining-Pothorst expedition would have had to be fitted out in Bergen, sail then to Iceland and then to Greenland for the start of the actual exploring voyage. Likewise Johannes Skolp's expedition in 1476 would be fitted out in Bergen, sail to Iceland and Greenland for pilots, and wait for the right time to start for the Northwest Passage.

In a letter dated July 14, 1493, written on behalf of the cartographer, Martin Behaim, to the King of Portugal, Dr. Monetarius mentions that the Duke of Moscow 'a few years ago had discovered the great island of Greenland, the coast of which stretches more than 300 leagues, and where there are still numerous colonies of the Duke's subjects'.[27]

This can only refer to the Johannes Skolp (Scolvus) expedition of 1476. The suggested 'Duke of Moscow' occurs, as Gustav Storm has pointed out, because several Latin documents state 'Johannes Scolvus Polonus' instead of 'pilotus'; Skolp was thus thought to be Polish

[27] G. E. Ravenstein: *Martin Behaim, His Life and Globe*, London 1908.

and Greenland thought to belong to Russia.[28] The significant thing here, however, is that there was still in 1493, a populous colony in Greenland, and Skolp had his headquarters there.

There is further evidence of the life of the Norse colony in Greenland in the Danish excavation of the cemetery at Herjolvsnes by Dr. Nörlund. The frozen earth had preserved the clothing used in Greenland, and some of it was of a kind which was designed in Europe after 1500. As it probably would take some years before this European style reached Greenland, we may assume that the Norsemen lived there until at least 1530-40.[29]

Several historians have contended that the Greenland colony disappeared sometime after 1410, for it was in that year that the last Knarr is reported to have come to Norway from Greenland.[30] Knarrs were the ships used by the King of Norway to fulfil his agreement to send at least two ships to Greenland each year. The report of the last Knarr more correctly suggests that after 1410 other types of ships were used for the Greenland trade. We can be sure that Pining and Pothorst did not sail in Knarrs, but had more modern ships.

The assumption of many that the Archbishop Erik Valkendorf collected sailing directions about 1520, in order to rediscover a defunct Greenland colony is also not reasonable. The Archbishop's intention was to help a neglected colony, not a dead one. He wished to increase Greenland's production and trade in order to better the income of the Church and toward this end he collected all kinds of information about Greenland, including sailing directions. His plan was to operate the colony as a private enterprise financed by the Greenland taxes for ten years.[31] Nothing came of this plan however, as the Archbishop fled to Italy with the onslaught of Lutheranism. No special sailing directions would have been needed for the Arch-

---

[3] Snorre Sturlason gives us our first reference to a family of the name of Skolp. In his sagas of King Inge (ch. 22) and of Magnus Erlingson (ch.6), Snorre tells us of the Skolp brothers, Simon and Jon, who were married to the daughters of King Harald Gille. The Skolp family lived in Halkjelsvik Volda, a few miles north of the peninsula of Stad which was the usual starting point on the west coast of Norway for journeys to the Faeroe Islands, Iceland and Greenland. Johannes Skolp, the leader of the 1476 expedition, was very likely of this family and thus a Norwegian. [29] A. W. Brögger: op. cit. pp. 166-75. [30] G. H. M. vol. III. pp. 165-75. [31] G. H. M. vol. III, pp. 482-87.

bishop's plan for the crossing from Iceland to the Greenland coast as the compass was now in use.

It was the outlawing of the Catholic Church in Norway in 1536, that brought about the collapse of the Greenland colony shortly after. For it was the Church that sent priests out to the far corners of its realm, sought to maintain and increase its subjects for the greater glorification of the Church, and therefore was interested in the people as Catholic citizens.

But as we have seen, though emigration decreased her population, the Greenland colony existed clear up to the Reformation, providing a definite link to the new lands across the sea. The knowledge of the Norsemen, far from being forgotten, was transmitted to the navigators of the world and not a few discoverers discovered lands the Norsemen had been using for 500 years and were still working. They had surely heard the stories of Vineland and some of them most surely had traveled to Iceland, Greenland and even the lands beyond

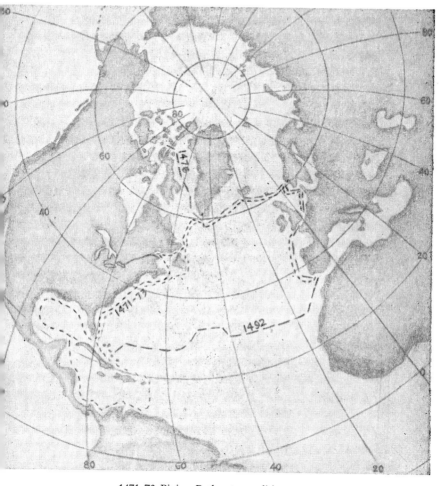

1471–73 Pining–Pothorst expedition
1476 Johannes Skolp expedition
1492 Columbus's voyage to West Indies

Columbus

Steure mutiger Segler! Es mag der Witz dich verhöhnen.
Und der Schiffer am Steuer senket die lässige Hand.
Immer, immer nach West! Dort *muss* die Küste sich zeigen,
Liegt sie doch deutlich und liegt schimmernd vor deinem Verstand.
Traue dem leitenden Gott, und folge dem schweigenden Weltmeer,
Wär' sie doch nicht, es stieg jetzt aus den Fluten empor.
Mit dem *Genius* steht die Natur in ewigem Bunde,
Was der Eine verspricht, leistet die andere gewiss.

(*Schiller*).

De muy pequeña edad entré en la mar navegando, é lo hé continuado hasta hoy. La mesma arte inclina à quien le prosigue à deséar de saber los secretos deste mundo.

The Pining-Pothorst Expedition

According to Carsten Grip, Pining and Pothorst were sent out with several ships 'at the request of his Royal Majesty of Portugal for the purpose of discovering new islands and lands in the North'. This indicates only that the expedition would have started from Greenland. Some writers have suggested that the expedition was dispatched to search for islands in the North Atlantic, but this could not have been the case.

Bjarne Herjulfsson sailed in 986 from Iceland to Nova Scotia; a few years later Eirik the Red and Thorstein Eiriksson spent a whole summer cruising between Greenland, Ireland, Iceland and Greenland. Different nations had sailed the North Atlantic for 500 years. The Icelanders and the Greenlanders knew in 1470 that there were no more islands to be discovered in these waters.

Gustav Storm and Fridtjof Nansen suggested that the Pining-Pothorst expedition went to East Greenland which both assumed was unknown at the time. We know now, however, that East Greenland was the first part of Greenland to be discovered. The northwest coast of Greenland as well as Baffin Bay and Baffin Land were also discovered by 1200,[32] and so the only uncertain territory lay southward along the American coast.

It was an opening to the west that Pining and Pothorst were to search for. It was the trade of the East that warranted a better route if one could be found.

Knowledge from the Norsemen

When King Kristjern I and his staff, at the request of the King of Portugal, started to plan an expedition for the purpose of finding a route to India and China, they had been collecting for years information about Helluland, Markland and Vineland. Pining and Pothorst had been on Greenland as naval officers as well as merchants;

[32] J. Kr. Tornöe: *Lysstreif...*, pp. 23–76.

they would have had the opportunity of discussing with the Greenlanders how much was known about the possibility of a route westward to China. By studying the narratives about the Vineland voyages, they could easily understand that there were three possibilities:

1. Between Helluland and Markland (through Hudson Strait and Hudson Bay)
2. Between Markland and Vineland (through the Gulf of St. Lawrence)
3. South around Vineland

The first possibility, the Hudson Strait and Bay, is very far north and there would be too much ice even in the summer for regular sailing. Moreover the Greenlanders knew the Hudson Bay and that there was no opening to the west. We must keep in mind that the Hudson Bay was the nearest and largest area for hunting of walrus for the tusks and supplying of wood for housing, fuel, etc., which had gone on for almost 500 years. They had named the region Marklands Botnar moreover; *botn* meaning 'bottom' with no other outlet, shows they knew there was no passage there.[33] So we may eliminate this possibility.

To India through the St. Lawrence

The Gulf of St. Lawrence, the second possibility, had many more potentialities than the Hudson Bay region. The Gulf is located on the fiftieth parallel and looked very favorable compared to the African coast where the Portuguese had tried for years to penetrate to India. Along a seaway on this latitude to the west coast of America and the Pacific Ocean, new islands and lands – almost anything – could be expected. No wonder that the King of Portugal wanted to know more about it and proposed to his friend, King Kristjern I,

---

[33] G. H. M. vol. III, p. 242, relates an account about the Northwest settlement (Norderseta): 'Höfðu þeir Norðrsetumenn sinar búðir eðr skála bæði i Greypum ok sumir í Kroksfjarðarheiði. Rekaviðir eru þar en eigi vaxinn viðr; tekr þessi norðskagi Grænlands helzt við trjám ok öllu hafreki, sem kemr úr Marklands botnum ...'. In translation: The hunters and trappers of the Northwest settlement of Greenland had their temporary or permanent houses in Greypum and Kroksfjardarheidi. Driftwood came there but there was no natural growth of woods; this northern point of Greenland received trees and all that drifted on the seas which came from Marklands Botnar (Hudson Bay).

to join in an expedition to investigate the territory. Certainly they had an agreement on the cost and the results of the expedition, a gentleman's agreement or a treaty, but no one has yet found it and it may be lost forever.

King Kristjern had another good reason for sending out such an expedition. His sovereignty over Markland and Vineland was as yet unchallenged in 1470. Vineland was considered by the Church to belong to the Kingdom of Norway since no one else had claimed these lands west of Greenland, and an expedition gave good opportunity to strengthen King Kristjern's sovereignty over Vineland as well as to explore the land.

The starting point of the expedition obviously had to be Greenland and the pilots Greenlanders along the coast of Labrador. According to the narrative of the Vineland voyages, the expedition would have to sail to the island north of the land (Scatari) and turn west, north of Nova Scotia, searching for an opening along the coast to the St. Lawrence River. Because of the enormous quantities of codfish they discovered from Labrador to Nova Scotia, the Portuguese called these lands Terra de Baccalao. Gaspar Fructuoso, too, mentioned in his story that Joao Vaz Cortereal had returned from Terra de Baccalao.

Coming to the St. Lawrence River, the expedition would discover there was no opening to the west and that they must return east to the Cabot Strait. But where would they go from there? They could return to Greenland, but it is not likely that a large joint Portuguese and Scandinavian expedition of two admirals and several ships would give up the search for a route to India so early. It was prepared for a long journey and as it was probably still early in the summer there was no reason not to continue south along the American coast. On the contrary, there was hope for an opening south of Vineland – the third alternative, which must have been considered most probable judging by the size of the expedition. Unfortunately we have no report from this expedition so we must infer its course by other means.

South, around Vineland

If the expedition had planned to sail south along the American coast to equatorial waters, it would have been an asset to have as advisers Portuguese sailors with experience from the African coast.

We have seen earlier that Sofus Larsen shows Joao Vaz Corte-Real to have been a member of this expedition. Joao Vaz could advise the Admirals how to provide food for the crew in the hot climate. And since the expedition was requested by the King of Portugal, it is likely that some Portuguese naturally participated in the journey.

The problems facing the Pining-Pothorst expedition would be the same as lay before Thorvald Eiriksson's crew which sailed south to go around Vineland.[34] We suppose King Kristjern and his staff were aware of Thorvald's journey described in the Flateyarbok and planned accordingly for a longer time for work south of Falmouth than the probable five months Thorvald's crew used sailing south and returning to Falmouth. Perhaps the Greenlanders could give more information about the journey in 1470 than the Flateyarbok gives now, and certainly they would have known something about Landa-Rolf's colonization of less than 200 years before. The American coast could not have been unknown to the expedition.

When the expedition passed Florida and sailed into the Gulf of Mexico and the coast headed north, they would believe that they had come to the west coast of Vineland. When the coast later headed west they would believe that they were still on the south coast of Vineland, but when it again headed south they would know that they had not come through and would have to continue further south and east searching for an opening to the Pacific. I would suspect that they searched as far as the Caribbean Sea where they would discover Cuba and the Bahamas.

If we should try to suggest a time-table for this expedition based on the information we do have, we would suggest that Joao Vaz Corte-Real and his party started from Portugal in March 1471, joining the expedition in Copenhagen in May, which fitted out in Bergen in June, and sailed to Iceland in July and to Greenland in September, 1471. It would not be appropriate to start from Greenland before late May 1472, reaching the Gulf of St. Lawrence in approximately the middle of July, and Florida in the middle of September, 1472.

Investigation in the Mexican Gulf and the Caribbean Sea would have taken some six months, to March 1473, before the expedition sailed back to Greenland in the middle of the summer of 1473 and

[34] J. Kr. Tornöe: *Early American History: Norsemen before Columbus*, Oslo 1964, pp. 78 ff.

Joao was back in Portugal in the autumn, which coincides logically with Gaspar Fructuoso's reference to Joao's journey to Terra de Baccalao.

We know that the expedition could not find an opening to the west, but the Admirals Pining and Pothorst had acquired a view of the entire east coast of North America; and so had Joao Vaz Corte-Real, and possibly other Portuguese who would bring this knowledge back with them to Portugal.

Another expedition

The most significant result of this expedition is the fact that the two Kings started to plan another expedition, and this time to the last hope – the Northwest Passage which had been known by the Greenlanders to exist, blocked by ice and uncertain as to where it led. We can be sure that the Kings would not have considered the difficult Northwest Passage if there had been the slightest hope of an opening farther south. I find this decision practical proof for my opinion that the Pining-Pothorst expedition had gone far enough south to determine that there was no opening to the west. Both the admirals and the Kings must have felt certain that the Northwest Passage was the last resort.

Again it looks as if the King of Portugal was the prime motivator. He called on Toscanelli, the famous astronomer and cosmographer, probably in the fall of 1473 and on June 25, 1474 the King received a letter and a map from Toscanelli which indicates the King had not given up his idea.[35] King Kristjern began to plan the Johannes Skolp expedition to the Northwest Passage which probably started from Bergen in 1476.

Sofus Larsen has erred in assuming the Johannes Skolp expedition to be the same as the Pining-Pothorst expedition and he did not know where they had gone.[36] It is necessary to make it clear that the Pining-Pothorst expedition was a great enterprise which lasted for at least two years. We are not sure just when the expedition actually started, but we know that Joao Vaz Corte-Real was back in Portugal in the autumn of 1473. On the other hand, all sources agree on the time 1476–77 for the Johannes Skolp expedition to the Northwest Passage.

[35] John Fiske: *The Discovery of America* vol I, London 1892. [36] Sofus Larsen: op. cit., pp. 89–90.

Thus there were two separate enterprises, and the second followed as a result of the failure of the first. If the Pining–Pothorst expedition had been successful and had found a route to the Pacific Ocean, there would not have been a Johannes Skolp expedition to the Northwest Passage.

The Northwest Passage

Centuries before the Johannes Skolp expedition the Northwest Passage had been discovered, probably by hunting parties from Greenland. The earliest reference to the Northwest Passage is found in *The King's Mirror* (Speculum Regale) written about the year 1250 by Archbishop Gunnarson of Norway, and the description of the water, which pours in through the sound from the outer ocean, indicates Jones Sound and Lancaster Sound.[37]

In the year 1141, the French historian Ordericus Vitalis had written that north of Greenland is no land[38] and before this Adam of Bremen, about 1070, had alluded to an arctic expedition of King Harald the Hard which is perhaps supported by accounts of dangerous streams and fogs encountered by King Harald related in the saga of St. Olav.

We know that the Norsemen did frequent the area far enough north on the west coast of Greenland to have also investigated the adjacent islands and waters leading westward. An investigation of an expedition of priests which started from Kroksfjordr in 1266, led me to the conclusion that Kroksfjordr is identical with Foulke Fjord at Smith Sound which separates Kroksfjordr from the region a little north of the Northwest Passage.[39]

In connection with my investigation Almar Naess computed the southern limit for the midnight sun on August 2nd, and found that the observation of the midnight sun could very well be made at Smith Sound, but at Disco, about 69 degrees north, the sun is far below the horizon on that date. Naess concluded that the fact that the priests saw the midnight sun cannot be ignored, and for that reason Nansen's suggestion that Kroksfjordr was at Disco Island will not fit in with the description of the observation of the midnight sun.[40]

[37] *G. H. M.* vol. III, pp. 336–40; J. Kr. Tornöe: *Lysstreif...*, p. 91. [38] *G. H. M.* vol. III, p. 428; J. Kr. Tornöe: *Lysstreif...*, pp. 85–92. [39] J. Kr. Tornöe: *Lysstreif...*, pp. 70–72. [40] Almar Naess: *Hvor lå Vinland*, Oslo 1954, pp. 126–27, 232.

A little further northeast of Foulke Fjord in Marshall Bay excavations by Erik Holtved produced some old camp sites and several implements of Norse origin indicative certainly of the existence of the Norsemen in the region.[41]

The saga of King Olaf the Holy contains a reference to Nánes as the northwestern corner of the Norwegian Realm. This area of Nánes is, in my opinion, identical with Cape Alexander in the Smith Sound region of Greenland.

The Johannes Skolp (Scolvus) expedition

We have seen that Dr. Monetarius mentions the Johannes Skolp expedition in a letter dated July 14, 1493. This is the earliest reference we have of the journey.

An English document from about 1575, printed by Nansen in *In Northern Mists* vol. II, p. 130, says: '... To find the passage from the North Sea (the Arctic Ocean) to the South Sea (the Pacific Ocean) we must sail to the 60th degree (Cape Farewell), that is, from 66 to 68 ... and this passage is called the Narrowe Sea or Streicte of the three Brethren; in which passage, at no tyme in the yere, is ise wonted to be found. The cause is the swifte ronnyng downe of sea into sea. In the north side of this passage, John Scolus, a pilot of Denmarke, was in anno 1476'.

There can be little doubt that John Scolus, by passing through Jones Sound or Lancaster Sound, attempted to sail from the Arctic Ocean to the Pacific Ocean. The information that the swift current from the Polar basin through the sounds carried the water into Baffin Bay is correct. The shallow water in the sounds keeps the ice back in the Polar basin creating ideal conditions for Arctic wild life to flourish there; no place in the Arctic could render a better sanctuary for whale, walrus, seal, polar bear and a variety of Arctic birds. The Narrowe Sea or Streicte of the three Brethren are probably old Norse names translated into English.

A globe made by Gemma Frisius and Gerhard Mercator in 1537 also shows the strait 'Fretum trium fratrum' which Johannes Scolvus penetrated about 1476. Various other sources exist referring to the Johannes Skolp expedition; and they are all agreed that the date of the event was 1476 to 1477.[42]

[41] Erik Holtved: 'Har nordboerne været i Thule distriktet?', *Fra Nationalmuseets Arbejdsmark* 1945, pp. 79–84. [42] Cf. Sofus Larsen: op. cit., pp. 84–90.

The King's Mirror

Of considerable interest is the book, *The King's Mirror* (Speculum Regale), which we have already had occasion to mention. Written by Einar Gunnarson, Archbishop of Norway about 1250, it contains a significant review of what was known and deduced at that time about the earth and the waters around it.

> Father. The matters about which you have now inquired I cannot wholly clear up for you, inasmuch as I have not yet found any one who has knowledge of the entire 'home-circle'[43] and its dimensions and who has explored the whole earth on all its sides, or the nature of the lands and the landmarks located there. If I had ever met such a man, one who had seen and examined these things, I should have been able to give you full information about them. But I can at least tell you what those men have conjectured who have formed the most reasonable opinions.
>
> The men who have written best concerning the nature of the earth, following the guidance of Isidore and other learned men,[44] state that there are certain zones on the heavens under which men cannot live. One is very hot and, because of the glowing heat which burns everything that comes beneath it, people cannot exist under this zone. It seems reasonable that this is the broad path of the sun, and I believe it is because this zone is pervaded with the sun's flaming rays that no one who wishes only a moderately warm dwelling place can live beneath it. These writers have also said concerning two other zones in the sky that under them too the land is uninhabitable; because, on account of their frigidity, it is no more comfortable to dwell under them than under the first mentioned where the heat is torrid. For there the cold has developed such a power that water casts aside its nature and turns into ice masses; in this way all those lands become ice-cold, and the seas too, that lie under either of these two zones. From this I conclude that there are five zones in the heavens: two under which the earth is habitable, and three under which it is uninhabitable.

---

[43] The 'home-circle' (kringla heimsins) was the Old Norse translation for the Latin orbis terrae, orb of the earth. [44] Isidore of Seville (d. 636) discusses the five zones of his Etymologiae, iii, c. xliv, xiii, c. vi, and in his DeNatura Rerum, c.x. The editors of the Soro-edition suggest that the 'other learned men' may be Macrobius and Martianus Capella, the famous encyclopedits of the fifth century (p. 195). But as these writers preceded Isidore by nearly two centuries, it is unlikely that their works were more than indirect sources for the scientific statements in the Speculum Regale. It is more probable that the reference is to such writers as Bede, Rabanus Maurus, and Honorius of Autun, though it is impossible to specify what authority was followed.

Now all the land that lies under the zones between the hot and the cold belts can be occupied; but it is likely that owing to location the lands differ somewhat, so that some are hotter than others; the hottest being those that are nearest the torrid belt. But lands that are cold, like ours, lie nearer the frigid zones, where the frost is able to use its chilling powers. Now in my opinion it seems most probable that the hot zone extends from east to west in a curved ring like a flaming girdle around the entire sphere. On the other hand, it is quite probable that the cold zones lie on the outer edges of the world to the north and south: and in case I have thought this out correctly, it is not unlikely that Greenland lies under the frigid belt; for most of those who have visited Greenland testify that there the cold has received its greatest strength. Moreover, both sea and land bear testimony in their very selves that there the frost and the overpowering cold have become dominant, for both are frozen and covered with ice in summer as well as in winter.

It has been stated as a fact that Greenland lies on the outermost edge of the earth toward the north; and I do not believe there is any land in the home-circle beyond Greenland, only the great ocean that runs around the earth. And we are told by men who are informed that alongside Greenland the channel is cut through which the wide ocean rushes into the gap that lies between the land masses and finally branches out into fjords and inlets which cut in between the lands wherever the sea is allowed to flow out upon the earth's surface.

You asked whether the sun shines in Greenland and whether there ever happens to be fair weather there as in other countries; and you shall know of a truth that the land has beautiful sunshine and is said to have a rather pleasant climate. The sun's course varies greatly, however; when winter is on, the night is almost continuous; but when it is summer, there is almost constant day. When the sun rises highest, it has abundant power to shine and give light, but very little to give warmth and heat; still, it has sufficient strength, where the ground is free from ice, to warm the soil so that the earth yields good and fragrant grass. Consequently, people may easily till the land where the frost leaves, but that is a very small part.

But as to that matter which you have often inquired about, what those lights can be which the Greenlanders call the northern lights, I have no clear knowledge. I have often met men who have spent a long time in Greenland, but they do not seem to know definitely what those lights are. However, it is true of that subject as of many others of which we have no sure knowledge, that thoughtful men will form opinions and conjectures about it and will make such guesses as seem reasonable and likely to be true. But these northern lights have this peculiar nature, that the darker the night is, the brighter they seem; and they always appear at night but never by day, – most frequently in the densest darkness and rarely by moonlight. In appearance they resemble a vast flame of fire viewed from a great distance.

It also looks as if sharp points were shot from this flame up into the sky; these are of uneven height . . .'

This treatise together with all the other information about the earth resulting from the voyages of the Norsemen and subsequent explorers constituted a wealth of information available to him who sought it. We have indicated the extent of that knowledge to about the end of the fifteenth century. Could Columbus have been one of those who availed himself of it? To what extent was he aware of the areas he was to discover?

The mystery of Columbus

In his *Historical Writings* John Fiske wrote a series in 1892 called *The Discovery of America*. In volume one of this work, pages 382–83, he says:

> About this time [between 1474–1480] Columbus was writing a treatise on 'the five habitable Zones', intended to refute the old notions about regions so fiery or so frozen as to be inaccessible to man. As this book is lost we know little or nothing of its views and speculations, but it appears that in writing it Columbus utilized sundry observations made by himself in long voyages into the torrid and arctic zones. He spent some time at the fortress of San Jorge de la Mina, on the gold coast, and made a study of that equinoctial climate.[45] This could not have been earlier than 1482, the year when the fortress was built. Five years before this he seems to have gone far in the opposite direction. In a fragment of a letter or diary, preserved by his son and by Las Casas, he says: 'In the month of February, 1477, I sailed a hundred leagues beyond the island of Thule, (to) an island of which the south part is in latitude 73, not 63, as some say; and it does not lie within Ptolemy's western boundary, but much farther west. And to this island, which is as big as England, the English go with their wares, especially from Bristol. When I was there the sea was not frozen. In some places the tide rose and fell twenty-six braccio. It is true that the Thule mentioned by Ptolemy lies where he says it does, and this by the moderns is called Frislanda'.

Commenting on this Fiske finds that 'taken as it stands this passage is so bewildering that we can hardly suppose it to have come in just this shape from the pen of Columbus . . . By Thule, Columbus doubtless means Iceland, which lies between latitudes 64° and 67°, and it looks as if he meant to say that he ran beyond it as far as the little

---

[45] *Vita Dell' Ammiraglio*, cap, iv; Las Casas, *Historia*, tom. i. p. 49.

island, just a hundred leagues from Iceland and in latitude 71°, since discovered by Jan Mayen in 1611...'.

In this suggestion I disagree with Dr. Fiske, as the ice condition, the long dark nights and the frequent stormy weather in February make it very unlikely that a ship at the time of Columbus could operate in these waters in that month. The waters concerned are located about 150 miles north of Jan Mayen. I worked for several years in the Norwegian Polar Institute and I have visited Jan Mayen and East Greenland two times, so I am familiar with this area. I find it much more likely that Columbus visited North Devon in the summer of 1477 in the Northwest Passage.

Charles Duff states his opinion in *The Truth about Columbus*, London 1957, pp. 32–33:

> If we admit that Columbus had become a man with a fixed idea in his mind – that of discovering the land to the West – in what direction would he be most likely to seek for some knowledge to verify his theories? It would be in the north, in Iceland or Scandinavia. He set out and to quote a later letter of his: 'In the month of February and in the year 1477 I navigated as far as the Island of Tile (Thule) a hundred leagues. And to this island, which is large as England, the English, especially those of Bristol, go with merchandise, and when I was there the sea was not frozen over, although there were high tides, so much so that in some parts the sea rose twenty-five fathoms and went down as much twice during the day.'
>
> To put together in some kind of logical order the events of this period of Columbus's life is like fitting together the pieces of a jigsaw puzzle. It is the conviction of the present writer, that no author has yet succeeded in piecing together the puzzle in such a way that it fits perfectly and produces a fully convincing picture.

Columbus at 73° north

Fiske's hypothesis that Columbus sailed to the 73rd degree north of Jan Mayen in February 1477 is implausible because, as we have noted, the weather and ice conditions in this area are prohibitive to a sailing ship reaching so far north. But Columbus could have sailed from Iceland to western Greenland in February before the field ice reached the west coast in the spring and summer.

As it happens Johannes Skolp took an expedition to latitude 73° – the Northwest Passage – just at this time. Columbus was interested in finding a new route to India and China and it would be likely that he would go to Scandinavia for information, and perhaps join an expedition like Skolp's. We know that the Portuguese had been in collaboration with the Scandinavians for fifty years already, and a Portuguese, Joao Vaz Corte-Real, had been on the Pining-Pothorst expedition. Still more likely, Columbus could have been in contact with the Portuguese King who engaged him as his observer on the expedition.

Upon the information we have, then, let us lay out a possible time-table for Columbus's journey to the Northwest Passage with Skolp, as we did for Corte-Real's journey with Pining-Pothorst.

Columbus left Lisbon in March 1476 after Corte-Real returned to Portugal and a new expedition was planned. He reached Copenhagen in late May, and Bergen, Norway, in the middle of June. Bergen was the center of the Greenland trade in Norway where it was easy to obtain information about Greenland, Helluland, Markland and Vineland from the sea captains as well as from merchant men. Communicating in Latin he would be able to gain much information from the bishop and the priests who probably had many books, including *The King's Mirror*, Vineland sagas and Ivar Bårdson's narrative from Greenland. Probably with a recommendation from the King of Portugal as well as from King Kristjern I, he would be able to obtain all possible good-will and help during the journey.

We may assume Johannes Skolp to be a shipowner, master of one of the ships which the King was obliged to send to Greenland every year. He probably took cargo to sell in the Faeroe Islands, in Iceland and Greenland, and started from Bergen for the Faeroe Islands late in July 1476. The Hauksbok gives this sailing direction for his route: 'From Stad is 7 days sailing to Horn on the east coast of Iceland. From Herdla [Bergen] the course is straight west to Hvarf in Greenland. In this course you steer so far north of Shetland that you barely can see the island in clear weather; but south of the Faeroes so far off that you can see only the upper half of the mountains over the horizon'. A look at a chart shows how correct the course is.

From Bergen to the Faeroes we estimate two or three days, and a week for trading. Now we remember the last sentence in Columbus's record: 'It is true that the Thule mentioned by Ptolemy lies where he says it does, and this by the moderns is called Frislanda'. Frislanda is the Faeroe Islands. When Columbus expressed himself so convincingly about the position of the Faeroes, it is reasonable to believe that he had stopped there and had the opportunity to observe the sun there. The indication here is that Columbus was a member of the 1476 Johannes Skolp expedition, for how else could Columbus have seen the Faeroes? By sailing from England or Ireland to Iceland he could not see the Faeroes.

The expedition would reach Iceland, which Fiske and Morison agree is Columbus's Thule, some time late in August or early in September, 1476. Johannes Skolp would stay in Iceland for some time, partly to collect information and partly to trade in different places. Columbus would have an opportunity to communicate with the two bishops and the priests and learn what information he could. In Iceland we know there were many books, including the Icelandic Annals.

It is logical that the expedition would have stopped in Iceland till about the end of February 1477, and then sail the hundred leagues to Greenland. Usually there is no field ice along the west coast of Greenland so early in the year which confirms Columbus's statement that the sea was not frozen. Johannes Skolp would have good opportunity to sell his cargo and unload his ship in the spring and fit it out for a voyage to the Northwest Passage in the summer, taking Greenlanders as advisers and pilots.

The southern coast of the island of North Devon is the north side of the Northwest Passage, which latitude is between the 73rd and 74th degrees. If Columbus was a member of the Johannes Skolp expedition, his observation could not be more exact. He emphasized that he sailed to an island in the 73rd degree, Lancaster Sound, and not in the 63rd degree, Hudson Strait. If we are not to reject Columbus's statements we must admit that the Northwest Passage is the only place which fits the details; and his date does coincide with the date of the Johannes Skolp expedition.

S. E. Morison has written in *Admiral of the Ocean Sea* vol. I, p. 33: 'The only incident in this Iceland voyage that sticks in one's crop is the 50 foot tides. Such can be found at only two or three places in the world; the spring range at Reykjavik is only 13 feet. It would be time and effort wasted to find an explanation of this'.

If Columbus had been in the Northwest Passage he would have seen the line of driftwood thrown up on land by northwesterly hurricanes in the winter. It is probably from this hurricane flood that

Columbus judged 26 braccio rise and fall of the tide and his statement would be quite understandable.[46]

Some historians have assumed that Columbus tried to give the latitude of Iceland as 73°. However, I am sure Columbus meant literally what he had written, that it was the island which he came to when he sailed beyond Iceland which was of latitude 73°. There is no reason to doubt Columbus's ability to make a correct observation of the sun at noon, and certainly we must assume that the master of the ship, Johannes Skolp, was an accurate observer and would make this observation for his report to the King. There can be no doubt that Johannes Skolp was in the Northwest Passage in 1476–77, whether Columbus was there or not. But Columbus emphasized that he was at the 73rd degree of latitude – Lancaster Sound, not Iceland, and I can see no reason to doubt his statement. Columbus's reference to the island where the English merchants go with their wares indicates he has interjected a statement about Iceland, a point which appears confusing here probably due to difficulties of translation of the original letter.

If Columbus had traveled to Iceland and Greenland, it would be inconceivable that he would not have obtained information about the new lands and the expeditions to them from the Vineland voyages to the Pining-Pothorst journey. Without doubt much of the information was also known in Portugal, and Columbus, being the interested party that he was, surely would have made himself aware of it. This would seem to indicate that Columbus journeyed to the West Indies intending to do so and in full knowledge of their existence.

The question I pose is this. Could the Portuguese reported to have informed Columbus of the Antilles have been a member of the Pining-Pothorst expedition? If so, then we have proof that the Pining-Pothorst expedition had reached the Antilles.

Were there any other Portuguese expeditions to the Antilles before Columbus? Not so far as we know. The Pining-Pothorst expedition appears to be the first, and it was jointly undertaken by the Scandinavians. As such expeditions were of prime interest to the King of Portugal we would assume he would have received thorough accounts of their outcomes. Therefore we hope that some written

---

[46] Otto Sverdrup: *Nyt Land*, vol. II, pp. 381–83; J. Kr. Tornöe: *Lysstreif...*, pp. 154–61.

report of the Pining-Pothorst and the Johannes Skolp expeditions is to be found somewhere, in Portugal or Spain, or perhaps in Italy, even Russia. Such a report would help us to clarify the history.

From the Pining-Pothorst expedition it is possible Columbus also received a map of the West Indies. An American named Robert Marx claimed recently that he knows Columbus had such a map when he sailed to the West Indies (*Aftenposten*, Oslo, March 28, 1964).

'Columbus was nothing but a good public relations man', says Marx, a professional deep-sea diver and navigator on board the Nina II, an exact replica of Columbus's ship, which sailed from Lisbon to the West Indies two years ago. In support of his Viking theory Marx had begun another journey which was subsequently abandoned. 'I have made certain investigations concerning Columbus during my stay in Europe, and I know he had a chart. The Vikings were there long before him. If I wrote an article on this, nobody would pay any attention to it. I have decided to show to the world that the Atlantic can be crossed under the same conditions as in the days of the Vikings and with the same kinds of ships as those that the Vikings had'.

Marx's certainty of a chart is interesting and if it is proven correct we would like to know more about the chart and ascertain its origin. In any case we certainly cannot regard Columbus as less than one of the greatest explorers of all time. For Columbus, in his age, was the man who had been farthest south as well as farthest north, on his journeys along the African coast in search of the route to the East and in his probable attempt to the Northwest Passage; it was he who knew more about the natural conditions on the globe than any other man of his time.

Further evidence of the kind discussed, documents, and other clues, are surely to be found. The outline of the events does not as yet present the full picture, and some of the evidence is circumstantial. There is much scope for further investigations, both in Europe and on the shores of North America, investigations which are necessary to confirm the unbroken historical chain that links the Vineland voyages to the explorations of Columbus and his contemporaries.

ADDENDUM

Early American History

When my previous book, *Norsemen before Columbus*, appeared, and after the present work had gone to press, Mr. Frederick J. Pohl wrote the following letter to the New York weekly journal *Nordisk Tidende*:

I must take exception to something in a recent book: "Early American History. Norsemen before Columbus", by J. Kr. Tornöe, Universitetsforlaget, printed (in English) in Norway by Harald Lyche & Co., Drammen 1964.

On page 21 Tornöe says: "Bjarni and his men had sailed for three doegr before they lost sight of Iceland. This indicates that they sailed northward along the coast from Eyrar, where Bjarni landed on coming from Norway, to Cape North, the most northerly promontory in Iceland. From there the distance to Greenland is shortest. This was not only the most natural and safe route to follow, but also the only course known at that time, because it was the one Eirik the Red took, and none but Eirik and his crew had at that time been as far west as western Greenland. At that time the Icelanders knew only what Eirik and his men had told them about the route to Greenland."

Tornöe forgets that Eirik the Red and his crew had sailed from western Greenland back to Iceland around Cape Farvel and from Cape Farvel directly across to Snaefellsnes, which was where the saga says they landed in Iceland. The reverse of this return voyage was obviously and undoubtedly what Eirik and his crew told folk in Iceland was the shortest route to Warf, the southern tip of Greenland. This unquestionably was the direction Eirik told prospective colonists to take since their destination was in the fjord on the western side of Greenland which he had chosen as suitable for settlement. This is what Icelanders knew before Bjarni arrived in 986 from

Norway. Because of his error of omission, Tornöe has Bjarni wandering off in the wrong direction, and all Tornöe's estimates of routes taken by viking explorers after Bjarni toward Vinland are unreliable because of this error which unfortunately in his case is basic to almost all his geographical conclusions.

My reply to him is as follows:

It is gratifying that an expert of Mr. Pohl's calibre has drawn attention to the question of Eirik the Red's route from Greenland to Iceland. This is one of the most important of the voyages which led to the discovery of America by Norsemen.

I take a special interest in discussing this question, because I myself followed roughly the same route as Eirik when I led the expedition on the Sealer *Signalhorn* to East Greenland in 1931 (see the map on page 55 of Meddelelser No. 56, S/S *Signalhorn*'s route). On this voyage I decided to attempt to establish Eirik the Red's route.

I wrote an article in *Norsk Geografisk Tidsskrift*, Vol. V, No. 7, 1935. This was the most thorough analysis of Eirik's route undertaken up to that time. The article was translated into English by members of the English Greenland Expedition, and partially incorporated in *The Geographical Journal* (Vol. LXXXIX, 1937, pp. 552–56).

The discorvery of Greenland and Eirik the Red's route

Islands Landnámabok and Björn Jonsson's *Grönlands Annaler* describe the discovery of Greenland roughly as follows:

"A Norwegian named Gunnbjörn, son of Ulv Kråka, sailed round Iceland when he came from Norway around A. D. 876. When he was out at sea to the westward he thought he saw a glacier to westward at the same time as he saw the Snæfell glacier in Iceland. They remembered this report in Iceland, and later Snæbjörn Galte voyaged out to live there. He discovered land and built a hut. But quarrels arose and killings took place; survivors forsook the land and went first to Hålogaland in Norway and later to Vadil in Iceland."

Around A. D. 985 Eirik the Red sailed from Iceland to Greenland. He sought for land in the same direction in which Gunnbjörn had seen it.

Björn Jonsson's *Grönlands Annaler* says (*Grönlands Historiske Mindesmærker*, I, p. 88):

"The driving force behind Eirik the Red's voyage to Greenland was simply the old folk's memory of the report that Gunnbjörn, son of

Ulv Kråka, thought he sighted a glacier to the westward at the same time as he could see Snæfell glacier in Iceland, when he was sailing west after leaving Gunnbjörn's Islands." (My translation.)

In *Grönlands Annaler, Grönlands Historiske Mindesmærker* I, p. 123, Björn Jonsson writes inter alia:

"It is said, furthermore, that the old priest Einar Snorrason, incumbent of Stadarstad, Oldhurygg, owned a large twelve-oared boat. She sailed out from Ondverdarnes with a cargo of dried fish which they wanted to take to market. The ship sailed out to sea so far that they saw both glaciers at the same time – just as Gunnbjörn had claimed – both Snæfell glacier and Blåserk on Greenland. Thus they had arrived in the neighborhood of Eirik the Red's route from Iceland to Greenland."

This indicates that Eirik the Red sailed from Iceland toward the high mountain on the Blosseville coast of Greenland. I asked some Arctic skippers if it was possible to see mountains of Iceland and Greenland simultaneously, and several of these confirmed that it was so. It seems that the old sources are reilable.

Again, it is said that Eirik arrived at Greenland at Midjökul, also known as Blåserk. I have maintained that Hvitserk and Blåserk are the two highest mountains on the Blosseville coast fo Greenland. They are the only ones that can be seen simultaneously with the Iceland mountains. Members of the English Expedition climbed Hvitserk and established its height as 12,200 feet, supporting my assumption that Hvitserk is the highest mountain on the Blosseville coast. It was from here that Eirik the Red sailed southwards along the coast to West Greenland. (See also *Geographical Journal* Vol. LXXXVIII 1936, pp. 193 and 214.)

In 1944 *Norges Svalbard- og Ishavsundersökelser* published my book *Lysstreif over Norgesveldets historie* as *Meddelelser* No. 56. In this book I have dealt in detail with the names mentioned on Eirik the Red's route (see pp. 8–38), and a thorough investigation is made of the discovery of Greenland and Eirik's voyage.

Thus I have not forgotten, as Mr. Pohl asserts, Eirik's return route from Greenland to Iceland. It is not written in the Saga, as Mr. Pohl maintains, that Eirik laid a course direct from Cape Farvel to Snæfellsnes. Indeed, Mr. Pohl seems to have misinterpreted or neglected the evidence of the Sagas.

In the *Flateyarbok* it says:
"Eptir um sumarit fór hann til Islands, ok kom i Breiðafjörð".
(Later in summer he sailed to Iceland and came to Breidafjord).

In the *Hauksbok* we find:
"En eptir um sumarit fór hann til Islands, ok kom i Breiðafjörð".
(And later in the summer he sailed to Iceland and came to Breidafjord.)

This indicates clearly that Eirik the Red did *not* sail direct from Cape Farvel to Iceland, as Mr. Pohl claims, but that he sailed from Cape Farvel along the east coast of Greenland to Angmagssalik, and thence, following the normal sailing ship procedure, on the latitude over to Breidafjord in Iceland. This was Eirik's first return route, and one that he could report on in Iceland. He could not have *reported* on a direct course. He could not even have given a compass bearing!

If Eirik had sailed direct from Cape Farvel to Iceland, his natural course would be to Reykjaness, which is the shortest distance away. But if he had laid this course Eirik would most probably have fetched up somewhere in southern Iceland east of Reykjaness because of the southerly current off Greneland, of which Eirik could hardly have been aware and therefore could not have taken into account.

But the most important indication that Eirik could not have taken the southerly route is that he had neither map nor compass, nor chronometer to fix his longitude. These things are difficult to comprehend for those who have no practical experience of sailing and navigation.

Many of the old sailing directions from Iceland to Greenland name Hvitserk as a mark. One should sail westward until Hvitserk lay to the north, and then alter course to the south-west (see *Meddelelser* No. 56, pp. 26–32). When sailing from Greenland to Iceland one probably took Ingolfsfjell, north of Angmagssalik, as a mark, since it is visible for 100 miles on the voyage east to Iceland.

Similar methods of laying courses were used on voyages between Greenland and America, as I have revealed in my original line of research in *Early American History* (see map p. 39). Researchers have had difficulty in understanding that these Norsemen sailed northwards along the west coast of Greenland when they really should have sailed southward to America. As I have shown, the reason why they did so was because they lacked maps, compass and chronometer. The most prudent and practical course was to follow the coastline as best they could and only turn toward the other land at the point where the crossing was shortest.

The Saga bears out my theory. Leif Eirikson crossed over to Baffin Land at a point so far north that there was no grass (*Early American History*, p. 53). Thorfinn Karlsefne sailed right up to Disco before crossing. His accounts of the physical features fit in with this route. but with no other; my theories seem to be borne out, not only on these points, but in the matter of the boundaries of Helluland, Markland and Vinland.

Bibliography

Ahlenius, Karl: *Olaus Magnus och hans framställning af Nordens geografi.* Uppsala 1895
Andersen, Magnus: *Vikingefærden.* Kristiania 1895
Andersen, O. Heitmann: *Det norske folks Busetning og Landnåm,* Oslo 1944
Anderson, Rasmus B.: *America not Discovered by Colombus.* Chicago 1874
Babcock, William H.: *Early Norse Visits to North America.* Smithsonian Miscellaneous Collections, Vol. 59 No. 19, Washington 1913
— Recent History and Present Status of the Vinland Problem. *Geographical Review,* Vol. 11, New York 1921, pp. 265—82
Bancroft, George: *History of the United States, from the discovery of the American continent to the present time.* Vol. 1, Boston 1834, 4th ed. 1838
Bang, Anton: *History of the United States.* Christiania 1863
Bárðarson, Ivar: *Det gamle Grönlands Beskrivelse.* Published from manuscript of Finnur Jónsson, Copenhagen 1930
Beamish, North Ludlow: *The Discovery by the Northmen.* Boston 1841
Berg, Henry: Vinland og Tidevannet. *Årbok* for 1955, Det Kongelige Norske Videnskabers Selskab Museet, Trondhjem 1956, pp. 45—65
Björnbo, Axel Anthon: *Adam af Bremens Nordensopfattelse.* Copenhagen 1910
— *Cartographia Groenlandica.* Meddelelser om Grönland, Vol. 48, Copenhagen 1911—12
Bobé Louis: Aktstykker til Oplysning om Grönlands Besejling 1521—1607, *Danske Magazin,* Ser. 5 Vol. 6, Copenhagen 1909, pp. 303—11
Bolton, Charles Knowles: *Terra Nova: the northeast coast of America before 1602.* Boston 1935
Brandes, Dr. Nanne: *Die grossen Ozeanbezwinger.* Bremen 1934
Brenner, Oscar: *Die ächte Karte des Olaus Magnus vom Jahre 1539 nach dem Exemplar der Münchener Staatsbibliotek.* Videnskabsselskabets Forhandlinger No. 15, Christiania 1886
Brögger, Anton Wilhelm: *Gamle Emigranter.* Oslo 1928
— *Ancient Emigrants.* Oxford 1929
— *Den norske bosetningen på Shetland, Orkenöyene.* Det norske Videnskaps-Akademis Skrifter, Hist. Filos. Kl. 1930, No. 3, Oslo 1930
— Den norske bosetningen på Færöyene. *Norsk Geografisk Tidsskrift,* Vol. 5 No. 6, Oslo 1935, pp. 321—33

— *Vinlandsferdene*. Oslo 1937

Bröndsted, Johannes: *Norsemen in North America before Columbus*. Copenhagen 1951

Bruun, Daniel: *Eirik den Röde*. Copenhagen 1915

Bugge, Alexander: Vore forfædres opdagelsesreiser i polaregnene. *Kringsjaa*, Vol. 11, Kristiania 1898, pp. 497—509

— Spörsmaalet om Vinland. *Maal og Minne*, Kristiania 1911, pp. 226—36

— Skibsfarten fra de ældste tider til omkring aar 1600. *Den Norske Sjöfarts Historie*, Vol. 1, Kristiania 1923, pp. 7—369

Bugge, Sophus: Hönen-Runerne fra Ringerike. *Norges Indskrifter med de yngre Runer*, Kristiania 1902

Daae, Ludvig: Didrik Pining. *Historisk Tidsskrift*, Ser. 2 Vol. 3, Kristiania 1882, pp. 233—45

— Mere om Didrik Pining. *Historisk Tidsskrift*, Ser. 3 Vol. 4 Kristiania 1882, pp. 233—45

DeCosta, Benjamin Franklin: *Notes of the Pre-Columbian discovery of America by the Northmen*. Charlestown 1869

— *The Pre-Columbian discovery of America by the Northmen with translations from the Icelandic sagas*. 2nd ed., Albany 1890

Delabarre, Edmund Burke: *Dighton Rock: A Study of the Written Rocks of New England*. New York 1928

DeRoo, P.: *History of America before Columbus, according to documents and approved authors*. Philadelphia 1900

Dieserud, Juul: Norse Discoveries in America. *Bulletin of the American Geographical Society*, Vol. 33 No. 1, New York 1901, pp. 1—18

Duff, Charles: *The Truth about Columbus*. London 1957

Espeland, Anton: Sjöfareren Didrik Pining. Norsk admiral, opdagelsesreisende og kaperförer. *Norges Sjöforsvar*, Årgang 2 No. 3, Oslo 1932, pp. 49—53

Falk, Hjalmar: *Altwestnordische Kleiderkunde mit besonderer Berücksichtigung der Terminologie*. Videnskabsselskabets Skrifter, Hist. Filos. Kl. 1918, No. 3, Kristiania 1919

Fernald, Merritt Lyndon: Notes on the plants of Wineland the Good. *Rhodora, Journal of the New England Botanical Club*, Vol. 12, Boston 1910

— The natural history of Ancient Vinland and its geographical significance. *Bulletin of the American Geographical Society*, Vol. 47 No. 9, New York 1915, pp. 686—87

Fischer, Josef: *Die Entdeckungen der Normannen in Amerika*. Freiburg i Br. 1902, English ed. 1903

Fiske, John: *The discovery of America*, Vols I—II. London 1892

— *Old Virginia and her neighbours*. Boston and New York 1897

Forster, J. Reinhold: *Geschichte der Entdeckungen und Schiffahrten im Norten*. Frankfurt 1784

Fossum, Andrew: *The Norse discovery of America*. Minneapolis 1918

Gathorne-Hardy, Geoffrey Malcolm: *The Norse Discoverers of America. The Wineland Sagas translated and discussed*. Oxford 1921

— A Recent Journey to Northern Labrador. *The Geographical Journal,* Vol. 59 No. 3, London 1922 pp. 153—67
Gebhardi, L. A.: *Kongeriget Norges historie,* Vols. I—II. Odense 1777—78
Geelmuyden, H.: Om gamle Kalendere, særlig Islændernes. *Naturen,* Aargang 7 No. 3, Kristiania 1883, pp. 38—43
Gini, Corrado: De norröne grönlandsbygders undergang. *Naturen,* årgang 81 No. 7, Bergen 1957, pp. 410—32
Gjerset, Knut: *History of the Norwegian People.* Vols. I—II, New York 1915, two vols. in one, New York 1927
Gosling, W. G.: *Labrador: its discovery, exploration and development.* London 1910
Gray, Edward F.: *Leiv Eriksson, discoverer of America, A. D. 1003.* London 1930
Greenland, Vols. I—III. Commission for the direction of the geological and geographical investigations in Greenland. Copenhagen 1928—29
Grönlands Historiske Mindesmærker, Vols. I—III. Copenhagen 1838—45
Harrisse, Henry: *The discovery of North America.* London and Paris 1892
— *John Cabot, the discoverer of North America, and Sebastian his son.* London 1896
Haskins, Charles Homer: *The Renaissance of the Twelfth Century.* Cambridge 1928
Haugen, Einar: *Voyages to Vinland. The First America Saga.* New York 1942
Hermansson, Halldór: The Vinland voyages: A few suggestions. *Geographical Review,* Vol. 17 No. 1, New York 1927, pp. 107—14
— *The problem of Wineland.* Islandica, Vol. 25. Ithaca 1936
Herteig, Asbjörn E.: The excavation of 'Bryggen', the old Hanseatic Wharf in Bergen. *Medieval Archeology,* Vol. III, 1959, pp. 177—86
Hertzberg, Ebbe: Nordboernes gamle Boldspil. *Historiske Skrifter* dedicated to Prof. Ludvig Daae on his 70th birthday, Dec. 7, Christiania 1904
Holand, Hjalmar R.: *The Kensington Stone. A Study in Pre-Columbian American History.* Ephraim 1932
— The 'Myth' of the Kensington Stone. *New England Quarterly,* Vol. 8, 1935, pp. 42—62
— *Westward from Vineland.* New York 1940
— *America 1355—1364.* New York 1946
— *Explorations in America before Columbus.* New York 1956
— *A Pre-Columbian Crusade to America.* New York 1962
Holm, Gustav: *Small additions to the Vinland problem. Meddelelser om* Grönland, Vol. 59, Copenhagen 1924
Holtved, Erik: Har nordboerne vært i Thule distriktet? *Fra Nationalmuseets Arbejdsmark,* Copenhagen 1945, pp. 79—84
Horsford, Eben Norton: *The problem of the Northmen.* A letter to Judge Daly, the President of the American Geographical Society. Cambridge 1889
— *The landfall of Leiv Erikson, A. D. 1000 and the site of his houses in Vineland.* Boston 1892

— *Discovery of American by Northmen.* (Adress at the unveiling of the statue of Leiv Eriksen delivered Oct. 29, 1887) Boston and New York 1888

Hovgaard, William: *The Voyages of the Norsemen to America.* New York 1914

Howley, M. F.: *Vinland vindicated.* Canadian Royal Society 1898

Humboldt, Alexander von: *Kritische Untersuchungen über die historische Entwickelung der geographischen Kentnisse von der neuen Welt.* Berlin 1836—52

Ingstad, Helge: *Landet under Leidarstjernen.* Oslo 1959

Irgens, O.: *Et Spörgsmaal vedkommende de gamle Nordmænds oversöiske Fart.* Bergens Historiske Forenings Skrifter No. 10, Bergen 1904

Jameson, John Franklin: *Original Narratives of American History.* Vol. I, New York 1905 (contains also Julius E. Olson's edition of the tales of the Vinland voyages)

Johnsen, Oscar Albert: *Noregsveldets undergang.* Kristiania 1924

Jones, Gwyn: *The Norse Atlantic Saga. Being the Norse Voyage of Discovery and Settlement to Iceland, Greenland, America.* London 1964

Jónsson, Arngrim: *Gröenlandia eller Historie av Grönland.* Copenhagen 1732

Jónsson, Finnur: Erik den rödes Saga og Vinland. *Historisk Tidsskrift,* Ser. 5 Vol. I, Kristiania 1912, pp. 116—47

— Opdagelsen af og Rejserne til Vinland. *Aarböger for nordisk Oldkyndighed og Historie.* Copenhagen 1915, pp. 205—21

— *Den oldnorske og oldislandske litteraturs historie.* 2nd. ed., Copenhagen 1920—24

— Flateyjarbók. *Aarböger for nordisk Oldkyndighed og Historie.* Copenhagen 1927, pp. 139—90

— (Ari Þorgilsson) *Islendingabok.* Tilegnet Islands Alting 930—1930. Dansk-Islandsk Forbundsfond, Copenhagen 1930

Kendrick, T. D.: *A History of the Vikings.* London 1930

Koht, Halvdan: The finding of America by the Norsemen. *Norwegian Trade Review,* Vol. 9 No. 3, Oslo 1926, pp. 37—43

— Review of Scisco: 'The Tradition of Hvittramanna-land' (Sagnet om Hvitramannaland). *Historisk Tidsskrift,* Ser. 4 Vol. 6, Kristiania 1910, pp. 132—36

— Norsk historieskrivning under kong Sverre, serskilt Sverre-Soga. *Edda,* Vol. 2, Kristiania 1914, pp. 67—102

Kolsrud, Oluf: Til Östgrönlands historie. *Norsk Geografisk Tidsskrift,* Vol. 5 No. 6, Oslo 1935, pp. 381—413

Kretschmer, Konrad: *Dei Entdeckung Amerikas in ihrer Bedeutung für die Geschichte des Weltbildes.* Berlin 1892

Larsen, Henry A.: *Henry med det store skipet.* Oslo 1964

Larsen, Sofus: Danmark og Portugal i 15de Aarhundrede. *Aarböger for nordisk Oldkyndighed og Historie.* Copenhagen 1919, pp. 236—312

— *Kilderne til Olaf Trygvasons Saga.* Copenhagen 1932

— *The discovery of North America twenty years before Columbus*. Copenhagen 1924
Larson, Laurence Marcellus: The Vinland Voyages. *The American Scandinavian Review*, Vol. 11 No. 9, New York, Sept. 1923, pp. 531—47
— The Kensington Rune Stone. *Minnesota History*, Vol. 17, Minnesota Historical Society, St. Paul 1936, pp. 20—37
Loffler, E.: *The Vineland-excursions of the ancient Scandinavians*. Amerikanistkongressens forhandlinger, Copenhagen 1883
— *The Vineland Excursions of the ancient Scandinavians*. Copenhagen 1894
Löberg, Leif: Norröne Amerikaferders Utstrekning. *Historisk Tidsskrift* Vol. 41, Oslo 1962, pp. 233—52
Magnússon, Finn: Om de Engelske Handel og Færd paa Island i det 15de Aarhundrede, især med Hensyn til Columbus's formeentlige Reise dertil i Aaret 1477, og hans Beretninger desangaaende. *Nordisk Tidsskrift for Oldkyndighed*, Vol. 2, Copenhagen 1833, pp. 112—69
Mathiassen, Therkel: *Skrælingene i Grønland. Grönlændernes Historie, belyst gjennem Udgravninger*. Copenhagen 1935
Merrill, William Stetson: The Vinland problem through four centuries. *The Catholic Historical Review*, Washington, April 1935
Mjelde, M. M.: Eyktarstadproblemet og Vinlandreisene. *Historisk Tidsskrift*, Ser. 5 Vol. 6, Oslo 1927, pp. 259—81, English summary
Morison, S. E.: *Admiral of the Ocean Sea*. Boston 1942
Moulton and Yates: *History of the State of New York*. New York 1824
Munch, Peter Andreas: Grönlands og Amerikas Opdagelse. *Almuevennen*. Aargang 2 Nos. 9—10, Christiania 1850, pp. 65—67
— *Det norske Folks Historie*. Christiania 1852—63
Næss, Almar: *Hvor lå Vinland*. Oslo 1954
Nansen, Fridtjof: *Nord i Taakeheimen*. Kristiania 1910
— *In Northern Mists*. London 1911
— The Norsemen in America. *The Scottish Geographical Magazine*, Vol. 27, Edinburgh 1911, pp. 617—32
Nielsen, Yngvar: Die ältesten Verbindungen zwischen Norwegen und Amerika. *Congrès international des Américanistes*. Vol. 14, Stuttgart 1906
— Nordmænd og Skrælinger i Vinland. *Det Norske Geografiske Selskabs Aarbog*, Vol. 16, Kristiania 1904—05, pp. 1—41. Also in *Historisk Tidsskrift*, Ser. 4 Vol. 3, Kristiania 1905, pp. 248—93
Nordal, Sigurður: *Orkneyinga Saga*. Samfund til Udgivelse af gammel nordisk Litteratur, Copenhagen 1913—16
Nordenskiöld, A. E.: *Om bröderna Zenos resor och de äldsta kartor öfver Norden*. Stockholm 1883
— *Bidrag til Nordens äldsta kartografi vid fyrahundra års festen till minne om nya världens upptäckt*. Svenska Sällskapet for antropologi och etnografi, Stockholm 1892
— *Periplus, an essay on early history of charts and sailing directions*. Stockholm 1897

Nordland, Odd: Øya med giftarmåls-vanskane. *Viking*, Vol. 17, Oslo 1953, pp. 87—107

Norges Gamle Love

Nörlund, Poul: *Buried Norsemen at Herjolfsnes. An archaeological and historical study.* Meddelelser om Grönland, Vol. 67, Copenhagen 1924
— *De gamle Nordbobygder ved Verdens Ende. Skildringer fra Grönlands Middelalder.* Copenhagen 1934

Nörlund, Poul and Roussell, Aage. *Norse Ruins at Garder.* Meddelelser om Grönland, Vol. 76, Copenhagen 1930

Nörlund, Poul and Stenberger, Mårten: *Brattahlid* (Researches into Norse Culture in Greenland). Meddelelser om Grönland. Vol. 88, Copenhagen 1934

Ólsen, Björn Magnússon: Landnamas oprindelige disposition. *Aarböger for nordisk Oldkyndighed og Historie,* Copenhagen 1920, pp. 283—300
— Landnama og Eiriks Saga Rauda. *Aarböger for nordisk Oldkyndighed og Historie,* Copenhagen 1920 pp. 301—07

Olsen, Julius E.: *The Northmen, Columbus and Cabot, 985—1503,* New York 1906

Packard, Alpheus Spring: *The Labrador Coast. A journal of two summer cruises to that region.* New York 1891

Peschel, Oscar: *Geschichte des Zeitalters der Entdeckungen.* Stuttgart 1858

Pinkerton, John: *Modern Geography.* Vols. I—II, London 1802

Pohl, Frederick J.: *The Lost Discovery. Uncovering the track of the Vikings in America.* New York 1952

Raestad, Arnold: *Kongens Strömme.* Kristiania 1912

Rafn, Carl Christian: *Antiquitates Americanae.* Copenhagen 1837, Supplement, Copenhagen 1841

Ravenstein, G. E.: *Martin Behaim, His Life and Globe.* London 1908

Reeves, Arthur Middleton: *The Finding of Wineland the Good. The History of the Icelandic Discovery of America.* London 1890

Robberstad, Knut: *Frå gamal og ny rett.* Vol. I, Oslo 1950

Roussell, Aage: *Norse Building Customs in the Scottish Isles.* Copenhagen and London 1934

Ruge, Sophus: *Geschichte des Zeitalters der Entdeckungen.* Berlin 1881

Schöning, Gerhardt: (Snorri Sturluson) *Heimskringla.* Copenhagen 1777
— *Norges Riges Historie,* Vols. I—III. Copenhagen 1771—73, 1781

Shetlig, Haakon: *Vikingeminner i Vest-Europa.* Oslo 1933
— Grönland og Vinland. *Det norske Folks liv og historie,* Vol. I, Oslo 1930, pp. 363—71

Skånland, Vegard: Supplerende og kritiske bemerkninger til Eirik Vandvik: *Latinske Dokument til Norsk Historie fram til år 1204. Historisk Tidsskrift,* Vol. 41 No. 2, Oslo 1961, pp. 136—38

Smith, Charles S.: The Vinland Voyages. *Bulletin of the American Geographical Society,* Vol. 24 No. 4, New York 1892, pp. 510—35

Söderberg, Sven: Vinland. *Snällposten,* Malmö Oct. 30, 1910

Sölver, Carl, V.: *Vestervejen, Om vikingernes sejlads.* Copenhagen 1954

Sprengel, Matthias Christian: *Geschichte der Europaer in Nord-Amerika.* Leipzig 1782
Steensby, H. P.: *The Norsemen's Route from Greenland to Wineland.* Meddelelser om Grönland, Vol. 56, Copenhagen 1917
Stefánsson, Vilhjalmur: *Greenland.* New York 1947
— *Northwest to Fortune.* London 1960
Storm, Gustav: *Eiriks Saga Rauda og Flatöbogens Gröenlendingapáttr sampt Uddrag fra Olafs saga Tryggvasonar.* Copenhagen 1891
— Om Zeniernes reiser. *Det norsk Geografisk Selskabs Aarbog.* Kristiania 1891, pp. 1—22
— Söfareren Johannes Scolvus og hans reise til Labrador eller Grönland. *Historisk Tidsskrift,* Ser. 2 Vol. 5, Kristiania 1886, pp. 385—400
— Om Betydningen af 'Eyktarstaðr' i Flatöbogens Beretning om Vinlandsreiserne. *Arkiv for nordisk filologi,* Vol. 3, Christiania 1886, pp. 121—31
— Studier over Vinlandsreiserne, Vinlands Geografi og Ethnografi. *Aarböger for nordisk Oldkyndighed og Historie.* Copenhagen 1887, pp. 293—372
— *Studies on the Vineland Voyages.* Christiania 1884—89
— Columbus på Island og vore forfædres opdagelser i det nordvestlige Atlanterhav. *Det norske Geografiske Selskabs Aarbog.* Kristiania 1893, pp. 67—85
— Review of Nordenskiöld: *Periplus. Nordisk tidskrift för vetenskap, konst och industri,* Stockholm 1899, pp. 157—61
Suhm, Peter Friderich: Forsög til en Afhandling om de danskes og norskes Handel og Seilads i den hedenske Tid. *Skrifter,* Det Köbenhavnske Selskab af Lærdoms og Videnskabers Elskere, Part 8, Copenhagen 1760, pp. 19—84
Sverdrup, Otto: *Nyt Land; Fire Aar i arktiske Egne.* Vols. I—II, Kristiania 1903
Swanton, John R.: *The Vinland Voyages.* Smithsonian Miscellaneous Collections, Vol. 107 No. 12, Washington 1947
Tanner, V.: De gamla nordbornas Helluland, Markland och Vinland. Ett försök att lokalisera Vinlands-resornas huvudetapper i de isländska sagorna. *Budkaveln,* No. 1, Åbo 1941
— Ruinerna på Sculpin Island (Kanayoktok) i Nain's Skärgård, Newfoundland—Labrador. Ett förmodat nordboviste från medeltiden, *Geografisk Tidsskrift,* Vol. 44, Copenhagen 1941, pp. 129—55
Thalbitzer, William: Skrælingerne i Markland og Grönland, deres Sprog og Nationalitet. *Oversigt over Det Kongelige Danske Videnskabernes Selskabs Forhandlinger 1905,* Copenhagen 1905—06, pp. 185—209
— *A phonetical study of the Eskimo language based on observations made on a journey in North Greenland 1900—1901.* Meddelelser om Grönland. Vol. 31, Copenhagen 1904
— *Four Skræling Words from Markland in the Saga of Eirik the Red.* London 1913

Thórðarson, Matthías: *Vínlandsferdinar*. Safn til sögu Islands og íslenzkra bókmenta, Vol. 6 No. 1, Reykjavik 1929
— *The Vinland Voyages*. Trans. Thorstina Jackson Walters. American Geograprical Society, Research Series No. 18, New York 1930
Thorkelsson, Thorkell: Den islandske Tidsregnings Udvikling, *Aarböger for nordisk Oldkyndighed og Historie*. Copenhagen 1936, pp. 46—70
Torfason, Thormod: *Historia Vinlandiae, History of Ancient Vinland*. New York 1891
Tornöe, J. Kr.: *Lysstreif over Noregsveldets Historie*. Norges Svalbard- og Ishavs-Undersøkelser, Meddelelser No. 56, with English summary, Oslo 1944
— Hvitserk og Blåserk. *Norsk Geografisk Tidsskrift*. Vol. 5 No. 7, Oslo 1935, pp. 429—43. Note from English translation by Michael Spender, *The Geographical Journal*, Vol. 89 No. 5, London 1937, pp. 552—56. See also Courtauld, A.: A Journey in Rasmussen Land. *The Geographical Journal*, Vol. 88 No. 3, London 1936, pp. 193—215
— Report on the expedition with the sealer Signalhorn to Eastern Greenland in the fall 1931. *Oslo Aftenavis*, No. 266, Nov. 18, 1931. See also *Aftenposten*, No. 418, Aug. 20, 1932
— *Early American History: Norsemen before Columbus*. Oslo 1964
— *Grönlandssaken i parti-politikkens tjeneste*. Oslo 1933
Vandvik, Eirik: *Latinske Dokument til Norsk Historie fram til år 1204*. Oslo 1959, p. 64 and pp. 170—72
Vartdal, Hroar: *Bibliographie des ouvrages norvégiens relatifs au Grœnland*. Skrifter om Svalbard og Ishavet, No. 54, Oslo 1935
Wheaton, H.: *History of the Northmen or Danes and Normans*. London 1831. French ed. Poul Guillot, Paris 1844
Winge, Herluf: *Grönlands Pattedyr*. Meddelelser om Grönland, Vol. 21, Copenhagen 1902
Wissler, Clark: *Archaeology of the Polar Eskimo*. Anthropological Papers of the American Museum of Natural History, New York 1918
Wormskiold, M.: Gammelt og Nyt om Grönlands, Vinlands og nogle fleer af Forfædrene kiendte Landes formeentlige Beliggende. *Det skandinaviske Litteraturselskabs Skrifter*. Copenhagen 1814, pp. 283—403
Wright, John Kirtland: *The geographical lore of the time of the Crusades. A study in the History of Mediaeval Science and tradition in Western Europe*. American Geographical Society, Research Series No. 15, New York 1925